The Video Demo Tape

The Video Demo Tape

How to Save Money Making a Tape That Gets You Work

Larry Benedict
Susan Benedict

Focal Press
Boston London

Focal Press is an imprint of Butterworth–Heinemann.

Recognizing the importance of preserving what has been written, it is the policy of Butterworth–Heinemann to have the books it publishes printed on acid-free paper, and we exert our best efforts to that end.

Library of Congress Cataloging-in-Publication Data

Benedict, Larry (Larry Charles)
The video demo tape : how to save money making a tape that gets you work / by Larry Benedict and Susan Benedict.
p. cm.
ISBN 0-240-80140-7 (paperback)
1. Video recordings—Production and direction. 2. Entertainers—Vocational guidance. 3. Motion picture industry—Vocational guidance. 4. Television broadcasting—Vocational guidance. 5. Video tape recorders and recording. I. Benedict, Susan (Susan Carol) II. Title.
PN1992.94.B46 1992
791.45′023—dc20 92-116
CIP

British Library Cataloguing in Publication Data

Benedict, Larry
Video Demo Tape: How to Save Money Making a Tape That Gets You Work
I. Title II. Benedict, Susan
778.59

ISBN 0-240-80140-7

Butterworth–Heinemann
80 Montvale Avenue
Stoneham, MA 02180

10 9 8 7 6 5 4 3 2 1

Printed in the United States of America

To all our friends and clients 1982–1989, and to those who lived on the shores of Angus Lake.

Contents

Preface

Today, the single most important tool for success in the entertainment industry is the demo tape. The first question asked by every prospective agent, director, or producer is, "Do you have tape of yourself?" Headshots are required, of course, but the real decisions are made on the efficacy of the demo tape. Competition in the entertainment industry is greater than in any other profession, yet the rewards for achievement can still reach fairy-tale proportions. Often, what stands between performers' goals and their jay job is a properly produced demo tape.

Although nothing guarantees instant success, the steps to an established career can be far fewer and more quickly and easily taken when the demo tape or video screen test has been assigned its proper value. In the eight years we owned and operated Angus Lake Productions, we saw it happen many times.

We became students of these tapes, analyzing and refining them to the limits of creativity. What do people look for in a tape? Exactly what turns them on and off? What will cause an agent to bring in a performer to read? What will cause a casting agent to recommend someone directly from their tape? The judgments needed when making a tape would regularly come down to a 30th of a second (one video frame) of screen time. We never considered a tape to be good enough until it was the best it could possibily be.

As actors ourselves, we took budgets very personally. It was part of our job, we believed, to invent every possible way to cut costs without sacrificing the tape's chances of getting the performer cast.

Where could you find the best blank tape for the cheapest price? What level of editing machine could do a broadcast job on an industrial budget? What effects would enhance a tape and what effects, regardless of their cost, would work against the performer?

In the world of business, how did what we learned from entertainment apply to the video resume, the product demo tape, and industry "pitch" tape?

How could we help nonperformers get their message across, even it if was their first time on video? What equipment was needed and what was simply overkill?

A million questions needed to be answered, and we were lucky enough to be in exactly the right place to find those answers. For eight years, in the heart of Hollywood, from early morning till after midnight, usually seven days a week, we videotaped and edited for performers and nonperformers in every imaginable field. Each one's sole purpose was to use videotape to advance his or her career.

We saw newcomers, who were so green (as we once were) that they needed help pronouncing the street names. There were old timers whose routines on black-and-white tape has us in stitches. There were swaypole artists, racehorse breeders, businesspeople, and psychics. There were so many actors, at every career stage, dealing with every career problem, that it would be impossible to remember them all. But each one had a story to tell and with every story came a lesson we could learn.

It is our belief that this book will show you how to make a demo tape that will get you work, how to do it right the first time, and how to spend only the money that actually needs to be spent. We also hope that by reading, you can share some of the wisdom, joy, the lessons, and the guideposts our clients so generously left with us, as they came through our studio making their tapes. During those eight years, their presence enriched our lives, and we hope that, in a smaller way perhaps, yours may be enriched as well. That is why we have dedicated our book to our clients.

1 Introduction

What is a demo tape? What does it look like, how is it created, and how good does it have to be to compete?

A demo tape is a commercial. It is your commercial. With it, you hope to convince the viewers that your product, whether it's your acting ability or a new welding torch you've invented, is the one the viewers should spend their money on. As with television commercials, your audience is sophisticated, overexposed, and has found it necessary to adopt a "show-me" attitude in simple self-defense. Nevertheless, they are looking for something special and when they find it, they will buy it.

What a demo tape looks like, how to make one that will compete, and how to do it within your budget is the subject of this book.

Just to give you an idea of where we're headed, we'll start by describing a tape we made for a motion picture production designer. Then we'll go back and begin at the beginning.

When the designer called, she explained her problem. Even though she had done thirteen movies and one television series, being considered for new work was always a gamble. Each of her projects had been designed in a completely different style. Although this was one of her talents, when it came to looking for work it was impossible to predict which movie would demonstrate her ability to create the look the producers wanted.

Another problem was that her tape would have to show sets, rooms, and locations: mostly master shots. When you string a bunch of master shots together, you have a very boring tape. What producers look for in a production designer is the ability to create sets and locations that enhance the emotional impact of their project. If we simply took shots of the different films and assembled them chronologically, the results would be static and unnecessarily empty.

We defined our objectives:

- We wanted the tape to elicit an emotional response from the viewer.
- It was important to demonstrate the designer's range, talent, artistic ability, creativity, and inventiveness.
- It had to be obvious that the designer has a lot of experience and can handle large projects.
- We wanted to generate a positive feeling about the designer herself.
- Our primary objective was to have prospective employers ask, after watching the tape, "Can you start Monday?"
- We wanted to do this in six minutes.

Given these objectives, we realized that "story" was of no interest and that chronology was unimportant. What is important for demo tapes besides displaying talent is creating an emotional response that encourages the viewer to hire you.

We studied the tapes for a week. Our client's job during that time was to find some emotionally strong music that she felt reflected her personality and tastes. When the right music was found, we laid it down on the tape. All the shots would be timed to the music, and appropriate dissolves would be used to enhance the mood.

As we became more familiar with the material, we began to notice patterns. Many shots included circles: a playground merry-go-round; a car wheel spinning in the snow; an athlete swinging around a high bar; a motorcyclist twirling his helmet in the air; the camera circling a room in a dolly shot; a skateboarder jumping and spinning. We decided to begin the demo reel with a montage that would preview the most exciting shots available. We then selected a series of themes for the six-minute tape:

- circles
- upscale professionals
- violence
- rooms
- candles
- lovers
- hallways
- art and painting
- victory

We selected only the shots that would contribute to the mood of each theme. For the circles grouping, we chose melancholy as the mood. The merry-go-round was a night shot; a man sits on its floor, turning it slowly with his foot as the painted horses come around into the frame. The athlete on the high bar was also shot at night; the bar is outdoors between two trees, and it is raining. The motorcyclist twirling the helmet is relishing the loneliness of a new romance. There's a sadness to the mood as the camera slowly scans the empty room.

Once we selected the circle shots that we felt met our criteria, we switched the mood to focus on upscale professionals. This sequence featured lots of bustle and activity, and success. Then came a section on violence. A fleeing man crashes into a fruit-seller's stall, spilling everything and falling to the ground. A boy in a wheelchair flies full tilt across a room and crashes against a door. A body is thrown from a roof and tumbles through sparking electric wires. Someone wields a knife on a stairway. A gang bursts through some garage doors.

The interesting rooms, all shot as masters—where the camera uses a wide angle to pick up the entire set—were carefully chosen for mood as well as design and variety. Where possible, people were included in the shot but with as little movement as possible. This series was done as one-second cuts. Next we used candles, a welding torch, a Christmas tree, a mirrored ball in a dance hall, light reflected in glass ornaments, and a grand chandelier.

For the segment on lovers, the mood was languid, flowing, private. We made the collection of shots as much a continuous pas de deux as we could. The hallways theme comprised mainly traveling shots, and we chose people moving briskly along interesting corridors. For art and painting, we chose the shots for taste and appropriateness to the style of the different productions. The shots we included ranged from fine art to posters to graffiti.

The final theme, victory, consisted of shots of people clinking glasses, shaking hands, laughing together, and hugging; a gymnast winning; guests applauding during a large dinner gathering; and a rock and roll band on a huge stage, their arms up in a victory salute. The final shot featured a ballerina in practice tights, sitting on a floor with her head down in what appears to be despondency, but suddenly she looks up with a broad victorious grin. Freeze frame, fade out, fade up name and "Production Designer" (white on black, four seconds), fade to black, music out.

Our client went directly from our studio to an interview for a new movie, and within hours after playing the tape, she was offered the job.

The Video Demo Tape Revolution

In the fifties, only the wealthiest of movie stars or their agents could afford the cost of producing their own screen test. Today, video screen tests are regularly requested, and it's the actor's responsibility to pay the bill. When Stanley Kubrick was casting *Full Metal Jacket*, he advertised for and received video screen tests of actors and would-be actors from all over the world. We saw many of these, and the production values ranged from fully professional to downright embarrassing. No one would think of sending a producer a snapshot instead of an 8″ × 10″ black-and-white glossy, and video screen tests should look just as professional. With competition as intense as it is, it is especially important to use all the available tools to your advantage.

Over the last decade, the video demo tape has revolutionized the hiring process in the entertainment industry. Led by actors, directors, and cinematographers, the reliance on the video resume to help land the next job has spread to all types of performers and to costume designers, stunt people, animal handlers, and even set builders and caterers. Producers and writers often go into "pitch" meetings armed with a video demo that they've created to enhance their presentation.

During the eight years that we operated Angus Lake Productions in Hollywood, we studied the video demo tape in depth. In that time, we created thousands of tapes for actors, performers, and other industry professionals. When a star was invited to promote his or her latest movie on late-night television, we were often the ones who edited the clip.

Large motion picture production houses and advertising agencies hired our company to produce their demonstration reels, which were once called *making of's*. This type of demo reel has now evolved into the electronic press kit (EPK), a far more elegant description of the demo reel at its most refined.

The use of videotape has spread well beyond the entertainment fields and into corporate America. Inventors, real estate agents, surgeons who develop new procedures, industrial companies that have created new techniques—in short, everyone with a service or a product to demonstrate—all are realizing the power of this visual medium, where the cost of production is so low that the video demo tape is within almost everyone's reach.

We have made reels for business people and educators. We assembled tapes for racehorse owners from videos of their horses' wins and comments from the trainers and jockeys. We shot video for breeders and made tapes of their yearlings, which they offered for sale each spring, so that prospective buyers could study horses of interest. Companies that wanted to land building contracts for shopping malls used our tapes to convince department stores to locate in their malls.

In all, we encountered almost every conceivable problem relating to demonstration reels and video resumes, and we dedicated ourselves to discovering effective solutions.

Now that many people in many professions are making tapes, almost everyone will be an actor at some point. Consider this: You're a floor manager in a large department store. The boss phones you one morning and warns you that he'll be coming by with a camera crew in an hour or so, and he'd like you to say a few words about the improvements you've made on your floor. "No big deal," he says. "It's just for the board of directors to look at." When you stammer that you get nervous in front of a camera and that you don't know what to say, he will probably reply "Oh, just tell it like it is. Relax. You'll be fine." Good advice, all right, but not easily followed.

When we say that nonactors will find themselves performing in front of a camera more frequently, we're not suggesting that you'll be learning lines. In fact, for a product or service demo, we seldom recommend memorizing a written script if you're not a trained professional. Nevertheless, when you give a presentation in front of a camera, however impromptu, you will employ the rules and techniques of the craft of acting. The more comfortable you are with these, the more successful your presentation will be.

The late George Britton, a Broadway star, shared the secrets of his commanding voice and terrific presence with business executives from many corporations, including the Ford Motor Company and Bell and Howell. Today, executives face another challenge. They must train themselves to feel confident, comfortable, and convincing in front of the ubiquitous video camera.

Much is going on in the area of video demo tapes, yet little information is available on how to make one. People must proceed almost completely uninformed, even though a mastery of this wonderful tool could lift their careers to new heights and save them a lot of money and frustration in the process. A well-made tape can put an actor out in front of the competitors, while a badly produced one might leave that actor toiling away at a day job. A variety of skills are required to bring a demo reel to successful completion. Although few of the concepts are difficult to master, the road is full of pitfalls if you're learning as you go. Suppose, for example, that you're shooting a product demo tape. The bright lights are on, and the boss is going over the presentation for the very last time. This is not the moment for the video department to be figuring out how to white balance the camera.

About This Book

We have divided this book into four parts: editing, shooting, technical considerations, and distribution. It may seem odd to discuss editing first and camera work second. There are two good reasons for our decision to do this. First, for many in the entertainment industry, the source material already exists. The problem is finding it, selecting what you want to use, and then editing it together. Sports personalities and media celebrities are in the same position.

The second reason for beginning with the editing section is based on the first demo tip, which follows. (Interspersed throughout the text, you will find short tips. These are thoughts and phrases that occurred to us in the heat of video battles and reminded us that our original intention had been to "drain the swamps," not to wrestle with electronic alligators. If you are using this book for reference in an actual shoot or editing session, the demo tips will function as guideposts to keep you going in the right direction.)

DEMO TIP: Shoot for the edit.

If you know how you're going to be editing before you do the shoot, you will save a considerable amount of time and frustration during both phases of the process. You can make intelligent, aesthetic decisions regarding close-ups, wide shots, and cut-aways. Books on cinematography devote whole sections to the cut-away, but for our purposes, a cut-away is a shot physically separate from the main action, yet directly related to it. It can be a shot of a hand dialing a phone or a car rounding a corner. (It may also be the reaction shot of another character. It can be used to condense time or comment on the main action.) When you shoot for the edit, you will know ahead of time whether you need to pan or to zoom. You won't end up shooting everything in sight; you'll just shoot what you're going to use. Once you've read the section on editing, you will be able to make these decisions in *preproduction*, the planning part, and execute them in *production*, the fun part, so that *postproduction*, the editing, does not become the slow part. As you will learn later, the more choices you must make in editing, the longer postproduction will take and the more it will cost.

We have tried to be as lighthearted as possible with the technical section. We became actors partly so we'd never have to think about all that left-brain hocus-pocus and got the surprise of our lives when it turned out to be fun! The reason for talking about it at all is now that we all need demo tapes, it makes us all instant producers. Producers have to deal with equipment and editing. They need to have at least some idea about what's happening technically. The more producers know, the more money they're going to save. Producers need to have control of the elements. (That's a little producer talk you can throw around.)

Even if you decide to hire someone to do the whole job, you'll soon realize that you know your product best (whether it's acting or creating designer bathtubs). You know exactly what you want your tape to do for you. So, all you accomplish by farming it out is to kick yourself upstairs. You've made yourself an *executive* producer, and you're still stuck with the very same decisions.

If you know the right word to use, if you can tell what's wrong with a tape by viewing it, and if you know that the session is slowing down and what to do about it, you will avoid a lot of wasted money and a lot of wasted time. It's as simple as that. In the chapters that follow, you'll find out how to interact with your editor, and you'll learn what the various machines do as well as a little about how they do it. We have been careful to be as technically clear and uncomplicated as possible while providing enough detail to satisfy those with a deeper interest in how video systems work. Nontechnical people might want to bite the bullet before booking a shoot or edit session and spend some time reading the few chapters in Part III. We structured this book around the questions we were most often asked while on the clock in edit sessions. By knowing the answers beforehand, you won't have to pay for that time.

In this book, we'll lead you through the process required for all the different kinds of tapes you might want to make. The actor's and performer's demo is probably the most complicated of all demo tapes, and we often use their problems and solutions as examples. The rules can be applied to all tapes, however, regardless of the type you plan to make. There are many specialized tapes, and individual sections are devoted to them. The suggestions, instruction, and editing assistance can be of use to anyone who must assemble a tape for promotional purposes, and the steps will guide you in turning out the finest product possible, both technically and aesthetically.

Some sections address the special situations of comedians, talk show celebrities, choreographers, news anchors, and other performers in active careers where video resumes are becoming essential. Some professions have yet to consider the usefulness of the video demo reel, and many could profit from an understanding of this persuasive medium.

Separating the process into its various components, we take you step by step from the assembly of your source material to the final distribution of your copies. Even if you choose to read only certain chapters, this book will help you save a lot of money on your first editing session and a tremendous amount over the course of your career.

What can you expect to gain from this book? To some extent, it depends on who you are. If you're an actor with screen credits, you'll learn where to find tapes of your material. If you're in college or regional theater and working toward a career in movies or television, you'll find out how to create a demo tape to get you in an agent's door without a lot of cost and struggle. If you've just been appointed to head the new video department at your company because the department was your suggestion, this book will explain the secrets of the video camera and show you how to shoot your surroundings in an efficient and flattering way.

If you're an editor or camera person who works on demo tapes or if you're wondering about getting into the field, you'll be interested in the sections of this book that describe the actor's needs and point of view. You'll discover how these tapes are structured and how they can achieve their goals. The demo tape and presentation reel have long been neglected while the "special event" arena (weddings and bar mitzvahs) is suffering from overcrowding. We've found our field to be fun, rewarding, and glamorous and well worth looking into by the industrial videographer.

If you are a facility owner or an edit bay designer who needs to put together a system that functions well for demo reels, you will discover clear descriptions of the machines that are required and how they are configured.

Have you ever wondered what everyday life was like for artists who were working their way up to stardom and the late-night talk-show circuit? Our work with demo tapes has taken us behind the scenes with thousands of working actors, the kind you've seen so many times but whose names you seldom learn. In this glimpse of their professional lives, you'll acquire helpful short cuts as you chart a course of your own.

Finding out about cameras, acting, lighting, editing, and distribution, as they apply to the demo tape is an easy and exciting introduction to the world of videotape production. An actor can learn about editing, a businessperson who's facing a camera can pick up acting tips, and an editor can gain an understanding of how an actor thinks and what an agent is hoping to see. Agents can help their clients make their video sales tool more effective, and artists can learn the agent's point of view. We've found every area of this business to be interesting, rewarding, and fun. It's our hope you'll experience some of the fun that we did as you learn to put together a demo tape that saves you money and gets you work.

Staying Calm

DEMO TIP: Don't overreact too soon.

The consternation and frustration experienced by people suddenly faced with the need to make a tape is widespread. Novices must face communication problems that develop when they try to explain themselves in a new language in a new electronic environment, often without even knowing which monitor they should be looking at. Time is often wasted running around town, and jobs have been lost because a tape wasn't where it needed to be or because the tape that arrived was the wrong one. Casting people have even received tapes that appeared to be blank, which embarrassed everyone and tossed the actor's chances out the window.

How could a tape appear to be blank? This is what happened. At our studio, we had just spent five hours making a tape for a very pleasant, easy-going, well-mannered young actor. The tape was fast paced and upbeat and showed an exceptionally wide range in just under six minutes. We made some dubs (checking each one to make sure everything had copied correctly), and the whole session had gone so well that everyone cheered at the final freeze and fade out.

The delighted actor took his tape straight to his agent, who was very happy with it and sent it over to the casting office by messenger. Exactly three hours after our victory celebration, the phone rang. A voice we hardly recognized was downright hysterical on the other end. "There's nothing on my tape! There's nothing on my tape!" screamed the formerly easy-going, well-mannered young actor.

"Hmmmm." We paused and reminded ourselves of the demo tip: Don't overreact too soon. Quietly we replied, "That's funny. We looked at all the copies before you left, remember?"

"Well, there's nothing there now, and they're going to cast somebody else!"

We know that it's very hard to think logically when something that bad happens. It's hard to think at all. "Did your agents look at your tape before they sent it out?"

"Yes."

"Did your agents rewind the tape before they sent it out?"

"Um . . ."

"Do you have the tape with you now?"

"Yes. I just got it home."

"Why don't you put the tape in your machine and try pressing rewind. We'll hold."

"Hang on." There was a pause, then the easy-going, well-mannered young actor came back on the line. "Uh, yeah, looks like that was it all right. Sorry I got so crazy."

"Don't worry. Glad we could help sort it out."

The very busy agent forgot to rewind the tape. The highly pressured casting agent saw nothing but snow and without the time to give it a second thought decided the wrong tape had been sent, pulled it, and went for the next one in the pile. Not wanting the highly pressured producer to start getting antsy, yet feeling downright embarrassed having given this actor a big build up, casting called the actor's agent and turned the phone lines blue. The agent then phoned the actor, and the actor then phoned us, and . . .

Need a happy ending? The actor got a second chance, the producer loved the tape, the actor worked on a major series and came back for the next episode as a guest star.

The solution is simple: Type "Please rewind tape" on the spine of each copy. (One of our clients thought typing "There will be a $1.00 rewind charge" would work even better.) When you call your agent's office to see how they liked the tape, just ask casually whether they rewound the tape after looking at it.

Not every problem is that easily resolved, but we hope that by sharing our experiences we will be able to assist everyone who has to work with demo tapes by providing logical answers to the multitude of questions that suddenly arise the moment you begin to make a demo reel.

Part I

Editing

2 Assembling Your Source Material

The tape that contains your unedited material is known as a *source tape*. It might have a show recorded off the air or all of your shooting footage. In an editing suite, any tape that is going to be edited is a source tape.

Each stage of an artist's career presents different problems in obtaining source material. For beginning actors without professional experience, the major problem is finding inexpensive ways to get yourself on videotape. Without source material of some kind, you can't assemble a demo reel, and without a demo reel, you can't get your foot in the door. Some beginning actors who are new in town risk their rent money to take an acting class, not so much to practice their craft but because they have heard that the classroom scenes are videotaped and they plan to get a demo that way. When they finally get their tape, they discover that the studio air conditioner is louder than their voice and that their face is inexplicably green.

More experienced actors face different problems: there's the "nurse" on "General Hospital" whose only line is "Good Morning, Doctor," and half of that is off camera; the child star who's no longer a child; and the mature actor who is finally able to return to acting after a sabbatical in real estate and now must locate old shows that are off the air and out of syndication.

Major stars and others with many years of work confront the problem of searching through hours of tape to select six minutes of their best work. A tape that's too long will never be fully seen, and one that's too short will fail to reveal the scope of their abilities. They also experience the psychological trauma of wondering why their agent is suddenly asking for a tape. Some experienced actors have never saved their work at all, and now they must track it down.

Why would actors whose careers have spanned decades and whose names are household words need tapes? These veterans frequently make this complaint: "My agent sent me over to make a tape. I've worked solid in this town for twenty years, and now they tell me I need an audition tape!" The actors feel insulted, and on the surface, they have a right to be.

But, consider this. You've had an account in your bank for twenty years. You go in one day, and the young clerk behind the plexiglass window refuses to conduct some simple transaction unless you show your identification. This is certainly an appropriate moment for your irate customer act. "I've been banking with this institution since they turned cameras with a crank," you say.

"Yes," says the kid. "But I've only been working here three weeks. I don't know anybody, yet."

Contrary to the popular myth, the most transient people in Hollywood are not the actors; they're the bank tellers and the agents. For whatever reason, the same people are never there when you go back. Without a tape, new agents may not know who you are. Allow us to offer the following stories.

One bright morning, long ago, we walked into an agent's office and met a young man who introduced himself as the new subagent. He had been given the job of sorting through a pile of head shots. As we chatted, it turned out that we had all been to the opening of *Arthur* the night before, and we raved about what a great film it was. Then he said, "There's an example of what I really love about Hollywood! Look at that old guy! I mean, he must have been kicking around Hollywood half his life, and suddenly he gets a part and he's a star!"

It took us a minute to fathom this. "Who?" Larry asked. "Do you mean John Gielgud?"

"Yeah, that's him! Isn't that great?" If Sir John's picture had come across this young agent's desk, it probably would have hit the wastebasket with the other unknowns. In all fairness, this young man went on to do very well and has learned almost every name in the business.

The next story is a short one. We know of someone making top decisions in the industry who once argued that Dick Van Dyke never had a television series.

It doesn't matter who you are. Tapes help the agents keep up with you, and it's no insult whatsoever to be asked to submit one.

The demo tape sometimes serves an additional purpose. One casting agent we know uses tapes as tie breakers. He narrows the candidates down to two or three, then everyone involved in the decision-making process watches the actors' tapes over and over, looking for a way to choose between talented equals. Presence, personality, command, style, attitude, bearing, consistency, versatility, experience, and that certain *je-ne-sais-quoi* are all scrutinized until one actor comes out the winner.

DEMO TIP: Collect all your material while it's current, and view it to make sure it has been recorded properly.

One of our clients allowed her mother to record all her shows for her for many years. Unfortunately, she recorded them at slow speed on an early Betamax, and the quality of the signal was so low that it couldn't be copied.

For Beginning Actors: Getting Yourself on Tape

Beginning actors who lack professional experience must be creative about finding ways to get themselves on videotape. Classroom scenes and stage plays that have been videotaped and video screen tests that you shot with friends are good places to begin, but you must use great care in selecting these scenes.

DEMO TIP: In nonbroadcast video, bad technical quality can make good acting appear amateurish.

Make sure that the technical quality is the best it can be. Don't invest an unreasonable amount of money in your early scenes, however, because you'll drop them from your tape as soon as you have real television or film experience. You will use your early video primarily to find an agent or to send to directors and producers who request video screen tests when conducting a talent search. You should send the highest quality tape you can *reasonably* produce.

Explore other sources for tape on yourself, including electronic schools and colleges that have camera students in need of the free services of a competent actor like yourself. Beginning actors work in return for a copy of the tape. If you've made it to the big city, you have some extra opportunities to get tape on yourself. In Hollywood, the American Film Institute (AFI) trains filmmakers and continuously casts talent willing to work for tape. Contact them right away. Besides the AFI, try the UCLA Film School, USC, and Cal State Northridge. Even if they are not casting nonstudents, you may find a film student working off campus who is willing to help you.

If you're performing in a regional theater or studying in a college theater program, check the Yellow Pages under "Video Production." Plenty of small video companies and even advertising agencies use actors. There could even be pay! Get in touch with local television and cable stations, and make friends there. That'll also be good practice for the big time. Friends don't necessarily help you get work, but they do help you get information. The more contacts you have, the more you're going to learn about what's going on. This is a social profession, and its product is communication.

For those working or learning in smaller cities or towns, many reasonably sized companies, even those that locate in rural areas, have video departments, and some larger ones have studios. Give them a call, find out who's in charge of video, and express an interest in their company. They might be intrigued enough by your call to find something for you to do. If they've never considered the idea of using video, team up with a friend who is handy with a camcorder, and pitch them on making a tape about their company. *To pitch* doesn't mean to come on strong, by the way. It means to present your idea in such an attractive, enthusiastic, and well-thought-out fashion that the company decides to buy it. Also *buy*, in this case, doesn't mean the exchange of currency. Don't expect pay for the first job; the company is doing you enough of a favor by being guinea pigs. This, in fact, is kind of an unwritten rule of the movie business. If you're a first-time screenwriter, it's hard to get as much as a dime for your first picture. It's not fair at all, but you have no clout without a track record. If the movie's a hit, the industry figures you'll make so much money on your second project that it will make up for the first. That's the theory, anyway.

If there's no industrial door to knock on where you are, there's probably a hospital. Nowadays, most have video departments that make tapes on training and medical procedure. If they knew where to find a little acting talent, particularly the free variety, they'd probably use it. If they accept your offer, you'll gain some practice at hospital-gown acting for that big break on the soap operas.

Those who are planning careers in news anchoring, news reporting, or sports-casting can follow the same trail. If you don't get anywhere, grab a friend, grab a camera, and shoot your own news show around town. Once you get good at it, you may even find an audience and a market. Who needs Hollywood or New York or Vancouver? We live in a global village, remember? Do it where you're living now.

Public Access

Public access television stations can be an excellent source of tape for novices, but there are some drawbacks. The worst is that practically all the elements are beyond your control. It's likely that everyone working on the set will be a student, and the time you are allowed will be severely limited. The waiting list for a time slot is sometimes impossibly long. Nevertheless, you can't beat the price, from $35 to free. If you hit the studio on the right day, the results can be excellent.

For cable or public access tapings, choose a scene or scenes that can be done on a talk show set, which is usually the most professional-looking set available. Avoid scenes that will put you in front of a blue cyc, or cyclorama. Borrowed from the proscenium theater, it is a background, often a cloth curtain, curved at both sides that can be lit to resemble the sky or other wide expanses. If it's not properly cared for, it can begin to hang in folds and it can even be possible to see underneath it. If your scene has a lot of movement, it's likely you'll end up in front of that draped material. Those curtains are no help at all, set-wise, when you're trying to convince the audience that you're a mafia don striding around your palace on the Hudson. (Follow the advice in Chapter 12 on shooting video screen tests and demo scenes.)

You can be quite inventive when it comes to ways to employ public access. In the early eighties, we worked for a group of comedy actors called "Imitation TV." They shot a half-hour show each month, sometimes in the public access studio and sometimes at a small professional studio in North Hollywood. Other times they went on location. They financed the show themselves, splitting the cost among the group and its producer, who was also perfecting her craft. They aired the show on the public access channel. Some of their skits and routines were eventually shown on national television, and the show got so popular that the cable station decided to run it twice a month. The actors got tape on themselves and lots of exposure by sending notices around to agents with the date, time, channel, and a little blurb about the success of the program.

After the show had aired, the actors called the agencies they had targeted to see if the agents had watched. If not, which was usually the case, the actors could respond with, "I know, it does air kind of late. I have a VHS of the show and would love to send it over. You might get a kick out of it if you have time to watch." (If you do something like this, remember that "no" is "no," and if you push agents too far they're liable to drive a stake through your 8 × 10 even as you talk on the phone.) You can send a little thank-you note to agents who take the time to listen to your calls. Then give them a rest until you've got something else cooking.

Talk shows on public access are always looking for subjects for interviews, and you might offer your services on the topic (what else?) an actor getting started in show business. The benefits will be twofold. First, it's a good chance to practice for the guest spots you'll be doing on national television when you've become a star. Second, the host might give you a neat little complimentary introduction that you can use to open your demo reel. If you use the intro on your tape, be sure it's good enough, and refrain from using more than a line or two from your actual interview. Agents don't consider a television interview to be acting, and they will wonder why interview material is included on your reel. You don't want an agent to wonder about anything except how they can get you more money.

For Experienced Actors: Collecting Your Source Material

For performers and artists who are already working in the entertainment industry, begin by collecting all the material that you have lying around the house. Then identify the shows that you don't have copies of and that you now need. Be sure to include your commercials. Demo tapes, like agencies, are categorized as *theatrical* (for television shows, movies, etc.) and *commercial*. You'll need to assemble your spots on your commercial reel, and they can also be used to enhance your theatrical reel in a variety of ways. Once you've identified the shows and commercials that you lack copies of, you're ready to begin your search.

When you start collecting your material from outside sources, try to get it on three-quarter-inch tape. This will provide you with the best picture and signal quality, and you will get the best results once the tape has been edited. We'll discuss tapestock in considerable detail throughout the book, but for now be aware that only certain stores stock this professional tape. You can find them in the Yellow Pages under "Video Tapes & Disks, Wholesale & Manufacturers." Most of the places listed will sell tapes individually. Be sure to compare prices because, as in every area of video, they vary widely for exactly the same thing.

The Production Office

To get a copy of a television show that you need, the first place to call is the production office of the company that made the show. Production offices usually charge a fee to make you a copy of the tape. It's a good idea to provide your own three-quarter-inch blank tape, if that is allowed, because studios sometimes run copies on well-used stock, which reduces the picture quality considerably. It's not a question of passing off old tape on the actor. Rather, the tape that is circulated inside the office is reused many times and they simply run your copy off on the same stock that they use for themselves. Do yourself a favor and give them a new tape to make your copy.

Some production offices will lend you a studio copy or director's copy with the requirement that it be returned within twenty-four hours. Be sure to respect this deadline because you'll want to borrow tapes from them in the future. They usually do not charge a fee for this service.

Many studios now provide copies for actors on video home system (VHS) or half-inch tape like the kind used in most home machines. While the quality is not quite as good as three-quarter, it is fine for your purposes. It's a good idea once again to provide your own new VHS stock to get the best quality. Refer to Chapter 3 to find out what type of tape to buy.

When the time comes to make your editing appointment, be sure to tell the facility personnel what formats your source material has been recorded on to make sure that their available machines are compatible with those formats. *Format* refers to the size or type of tape, such as VHS, three-quarter inch, and Beta. Many formats are currently in use, and many have the same name. Chapter 17 will help you sort them out.

Every major studio has its own policy for releasing tapes to performers, and the production office can give you the guidelines. As an example, here are some Hollywood policies. Offices in your area will operate in one of the following ways. Stephen Cannell has a very helpful one-day lending policy. Spelling-Goldberg does not lend or give out tapes under any conditions. CBS's "Young and the Restless" will lend you a tape if you call within three weeks after the show has aired, and "Days of Our Lives" runs copies for a fee. Other studios do not lend tapes directly to actors but will let you have a tape if your agent arranges it. All of these studios have good reasons for their policies, and it doesn't make them bad guys just because they don't lend out tapes.

At a certain point in your career, your agent can begin negotiating for a copy of your performance as part of your compensation. Although it's the surest way to get tape of your performance, your agent is the one who should decide when to start asking for this concession. You may have to give up something more important, such as money. For celebrities who do guest spots, it's best to ask for a clip of the show, or your segment, when you make the arrangements to be on the show. If your publicist makes your arrangements for you, he or she will probably take care of it, but it doesn't hurt to check.

Video Stores and Public Libraries

To obtain movies, some television specials, and even some out-of-syndication television series, your local video rental store is the most convenient source. Rental movies are on standard VHS tape and can be edited just fine.

Another source that is often overlooked is the public library. If you were in a how-to or other educational video, you'll probably have the best luck finding it at the library, and with your library card, the loan is free!

When picking up a tape from these sources, try to find the one that has sustained the least use and is the least likely to have a glitch in the best part of your scene or, as most often happens, over the title that you want to use. There's no harm in asking the rental clerk for a fairly new copy of the show. One of our clients searched out video rental houses that had recently opened for business. The copies he rented had probably never been played at all.

Some tapes contain copyguard to prevent piracy. Copyguard is a signal introduced into a portion of the video picture that is not apparent on viewing but that disrupts the picture when the tape is copied and makes it unwatchable. If the movie you rented flips or flickers, it will need to be edited with a time base corrector (TBC). This machine corrects the anomalies and signal errors that result from the instability of moving videotape being read by spinning video heads. The TBC reads the copyguard as an error and corrects it. We'll talk more about TBCs in Chapter 18.

Copyright. Actors and others are sometimes confused about the legality of using copyrighted material on demo tapes. The most recent ruling states that individuals may copy tapes for their own use as long as they are not exhibited for a fee. Because you are using unrelated segments of shows in which you appear, it is extremely unlikely that anyone will object. The strongest point in your favor, however, is that producers and casting agents could not function if actors were not editing their work onto the demo tapes that are used to make casting decisions. In all our years in Hollywood, we have never heard a single complaint about copyright infringement. If you are at all uncomfortable, you can call the producer and ask for a signed release.

Special Problems

Unaired episodes, unsold pilots, and out-of-syndication shows. When searching for unaired episodes or unsold pilots, the problem becomes more complex. Shows that are out of syndication or shows created by a production company that has been disbanded, as is often the case, present unique difficulties. Unsold pilots can be found during the summertime. They are aired by the network that commissioned them, or the production company that produced them, when time slots become available. If the pilot was "unsold," it means it wasn't picked up as a series by any of the networks. Sometimes these pilots were aired once; sometimes they never got off the shelf. Shows that are out of syndication can be found the same way. NBC used to run two unsold pilots back-to-back as a program called "Summer Playhouse." Most networks use these programs as fillers at some point because their production has already been paid for. Diligently scrutinize the television listings so that you can air check the show (record it off the air) when it appears. It may seem a little extreme to wait months until a show appears, but the time goes by faster than you think, and waiting is better than not getting the show at all. Don't delay making your tape if you need it soon; put together what you have, and update it when you finally have the show you're looking for.

If a series is canceled after three to five shows and your episode has not aired, you may be out of luck because it may never have been cut together. You can keep your eye on syndication, but the odds of finding the episode aren't too good. You could try getting to know the editor of the show and taking that person out for a very expensive lunch.

To pick up a rerun of a show you missed the first time around, note its original air date. If your show aired at the top of the fall season, it will most likely rerun during Christmas or at the top of the summer reruns. If the production office of this particular series is still open, they'll be able to tell you when your episode will air.

Syndication. Syndication works like this: A series is usually run for three seasons on a major network, most often the one that created it. After that time, it is often sold into syndication, meaning that independent stations buy the rights to air the show as repeats. Examples of independent stations in Los Angeles are Channels 13, 11, and 9, as opposed to ABC, NBC, CBS, and Fox. It's easy to discover the independents in your area by looking in the television listings. Shows are also sold into syndication to other countries, and if you're looking for something important, check with friends in Canada, Australia, or the United Kingdom. One of our clients had a very important tape that came apart due to a faulty cassette. We phoned Network 9 in Australia, where the show was running in syndication, and explained the predicament. They were happy to make us a dub from their master, which we received the next day, and were even kind enough to run it on NTSC (the American television system) for us so we didn't have to go to the trouble of getting it transferred.

Once a show goes into syndication, you can always find a rerun by discovering the time slot it's running in. Call the station or network, and speak to the programming department. They'll be able to help you as long as you've got the show number that is on your script or contract. They can give you a list of shows running but usually for only a few weeks at a time, so remember to be understanding and polite.

Commercials. Ad agencies are usually quite willing to help you obtain a copy of your commercial. When you call the agency, ask for the office of the person who was in charge of the campaign. The agency may run a copy of the spot or spots for you, but they often prefer to provide you with a copy of the schedule so you can air check the commercial yourself. You will get the date and the time of the spot right to the minute, usually for a week or more, so you'll have ample opportunity to record it. Getting your copy this way is preferable because you can control the quality of the tape on which it is recorded. If you call a service to air check your commercial for you, be sure to tell them that it's a spot they will be recording, and be very specific about the time it airs. In an effort to be helpful, service personnel sometimes start recording a show after the commercials have run, but if they know what you're after, they will start a little bit early so you have plenty of lead time before the spot appears.

If the ad agency runs a copy for you, they will generally charge for a three-quarter-inch copy but may sometimes provide a VHS copy for free. It's important to be extremely specific when requesting your spots. One client of ours ordered half a dozen spots from a campaign, and when he came in to edit, we discovered that he wasn't in any of them. They'd run all the spots with his female counterpart by mistake.

If you're looking for a commercial that is off the air or was made by an ad agency that is now out of business, check with MGS Services, a commercial clearinghouse in Chicago. They have a copy of every commercial that's ever been made. You can purchase a copy from them on three-quarter-inch tape. Their phone number is (312) 337-3761.

Very old shows are finding their way into specialty video stores, university libraries, and are sometimes advertised in nostalgia magazines. If you have friends who were on the show with you, get in touch with them; they very often have a copy of exactly what you need.

To view or transfer a tape that was recorded on a black-and-white, half-inch, reel-to-reel machine that predates cassette formats, try universities, art galleries, and museums. They are the most likely organizations to have the machine you need. There were few tape standards in those days, however, so even if you find a reel-to-reel machine it does not guarantee that your tape will play on the first one you find.

Transferring Film to Video

If you receive a copy of your work on 16mm film, or if some of your originals are on this format, many editing studios are able to transfer it to video for you. This should be done prior to your edit session.

The equipment used at this level is known as a film chain, and the resulting quality will vary according to the age of the machines in use. This system consists of a video camera and a film projector connected by a box that may contain mirrors, prisms, and focus rings, depending on the number of film sources it is equipped to handle.

More expensive units will be referred to as telecine, multiplexer, or by the brand name Rank-Cintel. The problem with transferring film to video is the difference in frame rates (film 24 frames and video 60 fields). This can be compensated for by the specialized telecine film projector.

If your film is on 35mm stock, you will need to have it transferred at a studio set up for this purpose. You will find studios listed in the Yellow Pages under Video Tape Duplicating and Transfer Services. These facilities are equipped to handle all film formats as well as slides.

Although home converters can be purchased for transferering Super 8mm film to videotape, they are not professional enough for our purposes and any Super 8 film footage you decide to use should be transferred professionally. It is important not to confuse Super 8 film of old, with the new home video formats, Super or S-VHS and 8mm video.

A Note on Courtesy

DEMO TIP: Be polite!

Whether you're borrowing tapes or trying to find out when a show will be aired, make those phone calls sparingly, and never pester. Telephone manners are quite important for success in your search. These searches are not only frustrating to the actor, they are also extremely important. If you're like most actors, you have waited until the last minute before doing a tape. As a result, emotions can run pretty high. If you annoy someone, it will do you more harm than good. It's important to remember that it's not the job of the people on the other end to provide you with the information you require, and when they drop whatever they are supposed to be doing to assist you, they are doing you a personal favor.

You need them. They do not need you. If you are open, pleasant, understanding, and friendly, you will get help. If your personal frustration seeps across the telephone line and they suspect it is directed at them, the help will end. Hollywood itself, like the movie business everywhere, happens on the telephone. Maintaining a pleasant phone voice is a good rule of thumb no matter what you're trying to accomplish.

As performers in Hollywood, we go out of our way to stay polite on the freeway no matter how egregiously we were cut off. Ever imagine giving some driver the gesture, only to walk into your audition and find that person sitting across the desk? Larry does recall a meeting where the opposite occurred: "I went into the interview and the casting agent said, 'Oh, was that you in the TR6 coming up Lankershim a little while ago?' 'Yup.' 'Neat car. We used to have one. We loved it. What year is yours?' The audition went especially well." Other cities may not be as centralized as Hollywood, but theatrical people everywhere tend to congregate in the same restaurants and to use the same services.

Source Material for Product or Service Demos

When making a product or service demo reel, source material can be assembled from footage you already have and from the new material you plan to shoot.

Earlier demo tapes, instructional material, photographs, and especially any exposure you've had on news or other television shows can all come in handy. It's important to make sure that the quality is consistent and that the footage is fairly recent. If the secretaries in the background are sporting beehive hairstyles, your material is probably too old. However, if you're planning to include a little video history of your company, the footage would be okay, perhaps even amusing.

During our years of work for Robert Abel of Abel Image Research (a winner of multiple Clio awards for special-effects commercials), he relied heavily on reworking existing footage when assembling his new pitch tapes or company demos. A pitch tape is simply a demo tape used to aid in selling an idea or bidding on a contract. For each new job he bid, he created a tape that specifically targeted the needs of that company or organization. (The parallel concept in television has been dubbed *narrowcasting.*) Product demos are discussed in detail in Chapter 11.

3 The Air Check

How did the air check get its name? We are happy to have uncovered the following explanation, which could actually be true. Back in the days of radio when the LP first came into general use, all-night disc jockeys were quick to discover that they could throw on an LP and head out for a bite to eat. The station managers were on their tail just about as quickly, and when passing the control booth on the way home, they muttered to the engineer, "Bill, run a tape on Ole' Sunnyboy Nightowl tonight for me, will ya? I want to check what he's doing on the air." In no time flat it was diminished to "Run an air check for me, will ya, Bill?" Finally, it became generic. To *air check* came to mean to record a show off the air for later review.

DEMO TIP: Your show is easy and inexpensive to tape when it airs, but it can be very difficult and costly after the fact.

When collecting your source material from current projects, you have two choices. You can call a company that offers an air check service, or you can record your air checks yourself. Whichever method you choose to record your material, you will need to keep a close eye on the scheduling of the episodes. They are not always run in the order in which they are shot, and a strong episode is often "plugged in" soon after the pilot to continue developing viewer interest. Be sure to keep your show numbers, and check with the scheduling department when necessary. *TV Guide* is good for planning, but for the very latest information, the daily paper is the most accurate source.

Another potential problem is the pre-emption. If a major sporting or political event is scheduled to precede the show you want, your show may be delayed or pre-empted altogether. Be careful that you (or the air check company you engaged) do not miss your performance because the preceding event ran overtime. This happens most often with basketball games. We have had clients who paid for huge ads in *Variety* stating the date and time of the show that they expected would catapult them to stardom, notified all the agents in town, and then were pre-empted. We know how tedious it can be to stay glued to air times, but a successful air check depends on it.

Always air check the entire show. This is especially true with soap operas. Trying to save a few pennies by ordering or setting up only the half hour that you think you're in can end in disaster. With soaps, a scene from the second half can easily end up in the teaser, and if you only record the second half of the show, you may not be happy with your decision. We will explain numerous ways to save money with your demo tape, but air checking only part of a show is not one of them.

Professionally Recorded Air Checks

Many video studios offer an air check service, and their prices vary widely. Find out where your friends have their tapes done and whether they are satisfied with the service. Then consult the various actors' and producers' guides for your city. They are generally published yearly and offer a list of establishments that specialize in doing air checks. The library usually has copies of these guides, and in Hollywood you can find them at French's bookstore or Larry Edmund's bookstore, among others. For other cities, check with the stores that specialize in theater books. You can also go through the Yellow Pages and the Recycler (in Los Angeles try under Video Services) to find an air check facility. The companies listed will not be the only ones that do air checks, but it is a convenient place to begin your comparison shopping.

Air check houses can be divided into two main categories. The first functions like a video clipping service; they record everything from early morning through 10 P.M., and you can call them up to three weeks after broadcast to obtain a clip of your performance. They generally edit out your scenes from the source tape and add the title if requested. You will want to request this additional service.

There are disadvantages to this type of service. First, it is expensive, generally costing just over one hundred dollars per show. Second, some companies record on VHS rather than three-quarter-inch tape. If you ask for three-quarter, they simply bump it up (transfer it from one format to another), defeating the purpose of getting the show on three-quarter in the first place. Because you will be editing and then copying your demo tape, generation loss is a serious concern. By getting your material on a bumped-up air check, you are down two generations before you even begin to edit. (Later we will discuss two new formats now coming into use: S-VHS and Hi-8. Both rival three-quarter-inch tape in quality and are much lower in cost. However, three-quarter and ordinary VHS are currently the standard for demo tapes.) Last, these services seldom offer sophisticated editing, and the facilities that do offer it may not allow you to be present.

The one indisputable advantage to the video clipping service is the ability to get your material after the fact. If your agent needs it right away or the lack of it is holding up your main editing session, bite the bullet, pay the price, and get on with your career.

The second category includes all the other editing facilities that offer air checks if you call them in advance. They range from the full-service facility to the industrial editing facility to the fellow out in the desert with an old hand-cranked editing system set up in a camper. His television receiver is outside the door on an oil drum. The latter sounds a little rustic, but the reception's probably pretty good. You most likely will be encountering the first two in the list and their advantages and disadvantages follow.

The Full-Service Editing Facility

Full-service editing facilities are large on-line editing houses. *On-line* means that the tape being made is the finished product that will go on the air or be packaged for distribution. A person making a demo tape should approach these facilities with great trepidation. Even if you can afford the breathtaking rates, your editing session will not last long enough to take advantage of the perks.

These facilities are set up to meet the stringent requirements of network broadcast television, and the difference in picture quality on your copies will be almost imperceptible compared to a properly maintained and fairly new industrial system. It is this second type of studio that is the most useful for making demo tapes.

Some full-service facilities do offer off-line bays at reduced rates, but compare prices carefully and inquire about the experience of the editor who will be assigned to your project. *Off-line* means that the tape being produced will not be the finished product and will be used as a sort of pattern for the on-line editing. If you make your demo tape on an off-line system and as far as you're concerned it is the finished product, then you may consider yourself to be working on-line. (Don't worry: By the time you finish this book you will have this concept down cold.)

The Industrial Editing Facility

The demo tape producer will be most comfortable in a reasonably priced industrial editing facility. It's not a Ferrari, but it will do almost everything a Ferrari does and much of it almost as well if you choose carefully.

The industrial editing facility is set up to service corporate educational and medical clients as well as demo producers of all types. In entertainment centers, many television producers have their favorite industrial studio where they can edit off-line at bargain rates. These facilities sometimes meet broadcast requirements and sometimes do not. A studio that doesn't meet broadcast standards may be more than adequate for your needs. Much on this subject is discussed in detail throughout the book.

Selecting a Studio

DEMO TIP: Decide on a studio before you order your air checks.

If possible, visit a few studios and select one for editing your demo tape *before* you book your air checks. It's far more convenient to do your editing where you have your air checks recorded; you will establish a relationship with the studio and hopefully with a specific editor.

Find a place that is relatively clean, quiet, and conducive to concentration. If unboxed tapes are scattered down an unswept hallway floor and a lot of screaming is going on, you may want to keep looking. The studio should appear to be established or at least doing some business. For example, the phone should ring a couple of times while you're there. The proprietors should be happy to show you tapes they've made and to discuss what they cost to make. Ask yourself whether the employees are friendly and helpful, and whether they pay more attention to the customer than to each other.

Ask about the studio's system for recording air checks. Check the picture quality of the air checks the studio produces. Are they recording off satellite, off air, or off cable? Satellite recording produces the best across-the-board picture quality, and if a facility has a satellite dish, that's a plus. Off-air recording is not necessarily bad, but you should look at tapes recorded from more than one channel to make sure the quality is consistent. In places around Los Angeles, where the transmitting towers are very close, off-air recording can be excellent, but Channel 2 is subject to little sparkles in the picture whenever the Santa Ana winds are blowing. Every major population center has its own broadcasting idiosyncrasies. Find out what they are before blaming a poor quality recording on the air check facility. Cable recording can be plagued by all sorts of interference due to the buried cable lines (a studio has little control over this), but cable should be available when you need it.

Air checks and edited masters (the tape on which your demo reel is initially made and which will be used to copy from) are handled in different ways by different facilities. There are three basic systems, although many variations are possible.

Some facilities expect you to own the air check, and air checking is performed as a separate service from the editing. Here you pay for the air check and the tape on which it's recorded. If you want to edit at that facility, you set up an appointment and return with your tape.

At other studios, you order the air check and pay for the service, but the actual recorded show belongs to the studio. This may appear to be less expensive than the first arrangement, but the studio only allows you to take your scenes off the air check, which remains their property. If you want to reedit at a future date, you may have to pay for the service again. Some of these places maintain permanent libraries, but you must edit at the facility to make use of their material.

Some of these establishments only release the whole show for you after they edit out the commercials. This means that the air check you take home is already down a generation, and beyond that, you are expected to pay for the time it takes to edit out the commercials. In this situation, your cost can go from what seemed like a reasonable rate (maybe even free) to over a hundred dollars. The advantage of the tape library could make the disadvantages acceptable, however.

The third type of facility records your air check, makes an editing appointment with you, then recycles the air check tape once your scenes have been edited. This is by far the most economical system and the one we recommend because you pay only for the service, not the air check tape. After editing, the expensive three-quarter-inch tape is recycled by the editing house. If you choose this system, be sure to record all your scenes onto your collected work tape, a tape that contains all of your performances assembled very simply and laid down chronologically. It is also known as a *compilation reel*. This is the tape you will use as your source for later reediting. From this point on, when you upgrade your tape, you'll have all your work to choose from, but you won't need to carry around all those source tapes and spend hours of time looking for the desired scenes.

All facilities charge their air check rate each time the air check is used because the rate is very low for the service they perform. The five or ten dollars saved by trying to split the same air check between actors is senseless economy. Air checking is a stressful job because someone has to be there making sure everything happens as planned, and when it doesn't, the operator has to get a backup working very quickly. Two people who press for a discount because they are both on the same air check do not endear themselves to the poor person who ran it. Remember that you will have many opportunities to save money while making your demo, and your editor is in a position to help you do it. If you do decide to team up with another actor on an air check, have it recorded at an establishment that runs the episode and sells you the tape outright.

When selecting a studio, another consideration is who will own the edited master. If you are charged for the edited master, the tape is yours. Like some photographers, certain editing studios believe they should retain control of the negs or edited masters. This means that you will be unable to change editing houses in the future should you wish to do so. The edited master is a very important piece of tape, and it should be under your control at all times. Once you have edited your tape, pay the studio for your master, and take it home with you. Be sure that your plans are clear to the facility before you go in to edit, and if you are at all uncomfortable, get your agreement in writing before you begin. If your editing session runs more than one day, always take your master home with you at night, and bring it back in the morning.

If your tape is already stored at an established facility you trust, there's no need to panic. Just make sure you know what the facility's policies are and that you're comfortable with them. Some places offer a master storage service, and one advantage of this is that it reduces your ability to access your master when you're tempted to hand it out after you've run out of copies. If a theatrical agency loses a master tape that you have paid hundreds of dollars for, you'll probably get so mad at them you'll end up having to find a new agent. This is one factor to consider when deciding what to do with your master tape.

Some companies serving the industry maintain vaults and charge to store precious film and tape in a fireproof, waterproof, dustproof environment. Companies that have just finished shooting a major project make protection copies of the originals and rush to store them all in a vault. They use working copies for off-line and do not go back to the originals until it's time to go on-line. You could store your tapes there too, but it would probably be cheaper in the long run to remake a missing master tape.

If the editing house has a policy of storing tapes so copies can be ordered by your agent, you can make a dub of the master and leave that dub with them. If the dub is lost or erased, it's a matter of a few dollars to replace it. If your edited master is lost, it's a matter of a few hundred.

As in any other business, you will encounter dubious deals and come-ons. If a studio advertises free air checks but their editing rates are higher than anyone else's, the air checks aren't free at all. Review all the costs before making a decision. Be alert to hidden charges and add-ons. Is it extra to put your name on the screen? Is it extra to shoot a head shot? What is the dubbing rate? Is there a high half-hour minimum rate? Is there a storage charge? The fewer extra charges there are, the more control you will have over the final cost.

What should you expect if an editing facility offers you a free consultation? It's not fair to ask the studio to help you design your tape for free. This is the part of the process that takes the most time and requires the most skill, talent, and experience. You should expect to be on the clock during that particularly creative session, even if no edits are being made. (Later chapters discuss tape design in great detail, and by following the methods described, you will save the most money in your edit session.) Studios use free consultations to introduce themselves to you and to provide a cost estimate. You will get some idea of their capabilities, and they will get a general idea of your source material. They may offer a suggestion or two about how they would approach your tape, and you should certainly feel free to ask any questions that come to mind. Just be aware that, like the free air check, the free consultation can mask some real expense down the line, so *caveat emptor* and all that.

In California, demo tape editing services are not taxable because a demo tape is used within the industry before the completion of the finished movie. It is only the end product on which a tax is charged to the producer or studio. This is not true for all states, however, and be sure to factor in any local taxes before totaling the final costs.

Missed Air Checks

You want to be as certain as possible that the air checks you book will not be missed. No human being or machine is infallible, though, so don't attack the studio with brands dipped in flaming oil if an accident does occur. One way to safeguard your ability to come up with a replacement for such an eventuality is to make a point of getting to know other actors on the set. Discuss air checking with them, and if something goes wrong, you can borrow someone else's copy. Not everyone can miss the same show at the same time, or can they?

When an air check has been missed for whatever reason, a good personal relationship with the editing house can be very helpful. Some places just apologize, and that's the end of it. No amount of righteous indignation will achieve anything but animosity. Air check facilities cannot guarantee air checks, but the better you known the people with whom you're working, the more you'll know about what to expect.

Some houses go to a great deal of trouble to help you find a missed show and are willing to utilize the methods outlined in Chapter 2. They will call the production office for you, for example. They may also phone around town to see if another establishment happened to record that show, and they may have a borrowing or duping arrangement with them. If you need the show immediately for a casting director or someone else, have the tape messengered over (at your expense because you will probably get a decent discount on your bill to help make up for the gaff), and continue with the editing session. If it's not that urgent, the house will be able to catch the show when it reruns, and they'll probably air check it for you for free.

Do-It-Yourself Air Checks

Many brave individuals assemble their source material by recording the shows on their home video cassette recorders (VCRs). Two of the advantages are obvious. It's cheaper, and you don't have to spend time picking up air checks from a video company.

Caution! If you are not already involved with video as a hobby and are not all that comfortable with the operation of a good-quality consumer VCR, this job may not be for you. Professional air check studios, though not infallible, earn their living by recording shows off the air and are far less likely to make a mistake. There are a number of ways to get it wrong, and any one of them will lose the show for you. Remember that a show is easy and inexpensive to tape when it airs, but it can be very difficult and costly after the fact.

However, with a little education, a straightforward approach to your machine, and a VCR of high recording quality, you'll save a bundle and have the convenience of a personal library of your work. Demos made from these tapes will certainly be acceptable, and if you follow the guidelines outlined in this chapter, your results can be excellent.

DEMO TIP: Do it yourself.

Never entrust this job to a friend or a relative, no matter how they rate their abilities. Something will always go wrong, and you won't discover the problem until you're in the editing studio and on the clock at fifty or one hundred dollars an hour. This is when you'll be faced with one or more of the following problems:

· The tape was recorded at the wrong speed. You'll have to pay for transfer time before the tape can be edited, and the picture you'll be left with will be grainy and barely watchable.

· The title of your movie of the week (MOW) is there with even the writer's credit, but following this is basketball.

· The heads on the machine were so dirty that the recorded image is 5 percent you and 95 percent snow.

· The VCR was tuned just slightly off the station, and sharing the screen with you are the ghosts of a wart hog family that drifted over from the PBS channel.

· In an effort to be helpful and edit out the commercials, your friend forgot to restart the machine until your big scene was over.

· The tape transport of the VCR was out of adjustment, and the tape can't be edited.

· The tape your economy-minded friend threw in to record your show was so old that it can't be edited.

· "Somebody" mislabeled your tape, and you and your editor get to sit and watch six episodes of black-and-white "I Love Lucy" reruns.

· The most important show just isn't in your box of tapes at all; no one knows why.

DEMO TIP: The more hands that your tape must pass through, the more likely it is that a mix-up will occur.

So, if you want a show taped and you want it done right, you have to do it yourself. After all, you have the most to lose if something goes wrong.

Recording Speeds

DEMO TIP: Always record on the standard speed—that is, the two-hour speed, not the four-, six-, or eight-hour speed—no matter what the salesperson told you.

Always record on the highest quality tape that you can, and always record on standard speed (SP). There are two very important reasons for recording on standard speed. To record six hours of information on a two-hour tape, three times as much information must be crammed onto the same amount of tape. This reduces picture quality considerably. If you're just recording an old movie so you can watch it later, it doesn't matter much. If you are going to be editing the tape and then making copies, you will be going down at least two generations, and by that time the picture will look terrible. (An editor would call it "marginal.") The second reason is that professional editing machines cannot work with slow-speed tape, and the tape will have to be transferred to three-quarter inch before it can be edited. Your picture will be degraded even further, and you'll pay more for the transfer than it would have cost to order an air check in the first place.

Buying Tape

DEMO TIP: For a show that you want to edit from, always use a new, name-brand, high-grade tape.

The tapes that do the best job are labeled "professional grade," "gold," or "special event." The box will be a little fancier than the other boxes of the same brand. In a manufacturer's line, the superlatives can often be confusing but the suggested retail price is the best comparison. The higher the price, the higher the grade of tape will be, even if the store is having a sale on one particular type because they're overstocked. If the manufacturer marks one tape at $13.99, another at $8.99, and a third at $3.49, this will give you a much better indication of tape quality than all the adjectives on the box.

Recently, manufacturers have been combining more than one grade of tape in a package of three or four. The high grade will be visible and advertised at a bargain price. When you get home, you will discover that you have one high-grade tape out of the four you need, and the three behind it will be the low-grade bargain quality. It's tiresome to have to check every single tape so fastidiously but it's a lot less annoying than returning to the store to try to exchange the unwanted tapes. It is not necessary to pay for hi-fi audio because regular television shows are not broadcast with hi-fi audio. You will find the best prices advertised in the weekend papers.

A new tape recently introduced to the VHS market uses an iron oxide known as *black magnetite*. These tapes are said to be superior to any tapes produced so far, but they may not be available at all stores. Major video discount stores will be your best bet until these tapes are in general use. Maxell calls its version Black Magnetite, HGX and 3M refer to theirs as Black Watch. There are a couple of versions of these tapes, and 3M's very best is called Black Watch, Hi-Pro VHS.

Fugi's finest VHS tape, although not black magnetite, performs exceptionally because of its double coating, and Sony's V HG is their top-of-the-line offering in the pro or special-event category. For those who have invested in Hi8 technology, Metal Evaporated tape, designated ME, rather than Metal Particulated tape, designated MP, will give you the best results, assuming the considerable difference in price is acceptable.

Avoid inexpensive tapes, and don't be fooled by the packaging. Some cheap tapes often look almost like expensive brands, and they are sometimes called something close to a popular brand. You'll see things like Maxwell instead of Maxell. These tapes are very cheaply made and have a thin oxide coating, a low-grade adhesive, stretchable plastic, and poor quality control, allowing the tape to come apart in your machine without notice. These tapes make the heads dirty, cause smeary colors, and have poor resolution. The signal is often so low that it can't be edited accurately. For your purposes, cheap tape is the most expensive tape you can buy.

Buying a VCR

DEMO TIP: Invest in a VCR with the best recording quality that you can afford.

Obviously, the VCR you buy will have a great effect on the success of your taping sessions and on your level of satisfaction. There are a few things to keep in mind when selecting the best machine for your purposes. The VCR should be a well-known brand and should have the highest advertised resolution. Resolution can range from 240 lines for an inexpensive VHS to over 400 lines for an S-VHS or a Hi-8. If you buy an S-VHS or a Hi-8, make sure that your studio can work with that format.

In terms of features on the VCR, you'll want digital tracking, flying erase heads (found on all the better machines) and a cable-ready tuner. The machine you choose should be programmable at least two weeks in advance or as long as you plan to be away. Two weeks is usually sufficient. Editing capabilities are not worth the extra cost unless you're thinking of becoming involved with home video as a hobby. A jog shuttle is very convenient if you can afford it. This little wheel on the front of the VCR allows you to review a tape at varying speeds.

Finally, a counter that reads in hours, minutes, and seconds is extremely helpful when you log your tapes and even more helpful in your edit session. Unfortunately, the prices for machines with this feature start at more than four hundred dollars and can go as high as twelve hundred dollars even in the largest discount stores, which generally have the best deals. However, as a professional, everything you spend on VCRs is tax-deductible.

When you purchase your VCR, request one that is new in the box, that is, one that's never been opened. Before your leave the store, check the box; make sure that the staples are all intact and that the shipping tape that seals the top, sides, and bottom is all the same. This way you'll be certain you're not getting a machine that's been returned for some reason. In the video business, boxes that have been opened qualify as B-stock. If you elect to go with an opened box, make sure you're getting some kind of discount.

Once you've decided to purchase a particular machine, you can ask the salesperson to open it up and check it over with you. Make sure all the parts appear to be there and are properly sealed in plastic with the cables neatly wrapped. It's very important to shop at a store that has a thirty-day, no-questions-asked return policy.

When you get home, open up the manual and turn to the page that lists all the included parts, cables, and accessories to double-check that they are all there. Now you can take your time in a more relaxed atmosphere. If it has dials, see that they are in the detente, or centered, position and that the little white marks are not pointing every which way. Check around for finger smudges and other signs of use. If you come across a little of this, there's no need to panic. Just be especially attentive when you run the machine through its functions. Very few VCRs fail during regular use. The majority that do, fail in the early days of operation, so learn to use your VCR as soon as you can, and use it a lot within the first thirty days.

Extended service plans. Extended service plans (ESPs) cost almost as much as paying for a repair, and they're primarily a way for the store to earn more on the sale. By the time the machine is worn enough to need the plan, it will be so old that you'll want a new one. If you follow the steps outlined above when you buy the machine, you can be confident in it. Even though most ESPs do what they promise and are great for peace of mind, we usually pass on them and take a chance on having to pay for a repair down the line. It is important to note that many stores rely on outside companies to provide the extended service plans and actually purchase these plans from the subcontractors before selling them to you. If the extended service plan company fails, as many have recently, the store is not responsible and the money you paid for your ESP is lost.

Catalog shopping for a machine is fine, especially with National Services such as American Express, Citibank, and the others. There are two drawbacks, however. The prices are much closer to suggested list than your local discount store and the models offered are not necessarily the latest, so it still pays to shop around before you order.

Learning to Operate Your VCR

DEMO TIP: Study the manual in detail, and practice on television programs that don't matter.

If you can follow a recipe in a cookbook, you can understand and operate your VCR. Set aside an entire day, and follow your manual step by step. We're old hands at reading manuals, but we still use a trick we used at the beginning. We follow every word by moving a finger along each instruction. We make sure we have read each word and know what it means before we move on to the next one. If you find this a little embarrassing, don't do it in front of others. By the time you finish, your command of the machine will impress the pants off your friends anyway.

Work with the VCR as you study the manual, and don't hurry. You've already set aside the time, and if you need more, schedule it for yourself. Make sure you can program the VCR in your sleep. If the machine doesn't do what you want it to right off the bat, be secure in the knowledge that most new machines are trouble-free, and figure out what you might be doing incorrectly. We have two little sayings that have kept our sanity more or less intact, and they are appropriate when you're getting to know your new machine: "Don't overreact too soon" and "It's not an airplane; it can't crash." Frustration is your enemy, Zen is your friend, and always let your finger do the reading.

Now that you can program your VCR in your sleep, don't do it. Make sure that you're fully conscious, that you're not thinking about anything else, and that no one is trying to talk to you. Program your start time, program your stop time, and think of nothing extraneous in between. Finally, make sure the little speed switch is on "SP." That's all there is to it.

Recently, a device went on the market (VCR Plus by Gemstar) that looks like a remote control but bypasses all the programming functions of the recorder. You can select a show and record it simply by typing in the five-digit number that appears next to the show in the TV listings. To schedule a recording, type in the number, and this little gadget starts your machine at exactly the right moment and turns it off when the show is over. That is all there is to it. It even switches channels all by itself. It can be programmed to record one show at a time or to record the same show all week. It works with any VCR that has a remote control, no matter what the brand. Not all shows list the five-digit numbers, however, particularly on weekends, so you're not off the programming hook completely. However, recording at home now makes a lot more sense.

4 Selecting and Preparing Your Source Material

We conducted the world's shortest interview with a Hollywood casting agent.

AUTHORS: What do you look for in an actor's demo tape?
CASTING AGENT: Talent.

Boy, those people don't mince words. When you begin to select and prepare your source material, keep that word in mind. Every so often as you watch yourself on the screen, cast yourself in the role of a casting agent. View the tape as if you were sitting in the busy casting room with all those people, sifting through tapes and looking for someone with talent to fit a specific part.

What do casting agents want? They want to be hired again by the production company. They want people to ask, "Where did you find those actors? They were great!" They want to be able to sleep nights between the time they go out on a limb to sell the director on you for the part and the time the project is shot. Above all, they want to help create a superior film or show.

Casting agents bet that they can make a living by picking the right people to cast in shows. They look for people who don't start making trouble the minute they get to the set. They want to cast people who can hit their marks and remember their lines. They want to cast people with talent. The more of these needs your tape can satisfy, the more often you'll be called.

Selecting Your Best Material

Dealing with Profanity and Nudity

Try to omit profanity from the tape. If it's embedded in the story and an integral part of the character, fine. However, if the profanity goes on and on, is unsupported, or seems to come out of the blue, avoid it. Use a different scene or cut around most of it. The expletives can be jarring and can cause people to stop thinking about your talent and start noticing the gratuitous use of profanity instead. This is particularly true of a screen test that you shot yourself. When you are looking for a scene to use for your screen test, why lock yourself into a character who uses profanity for no good reason? Just pick another scene. Never solve a problem when you can avoid it.

Nudity can also be distracting. Like profanity, it makes the viewer stop thinking about the acting. If your tapes are being watched for television and mainstream casting, eliminate any nude scenes if possible. The producers and directors of well-made feature films that contain nudity go to a lot of trouble to structure these scenes within the context of the story, but on your demo tape all that support is gone. There is little time to include redeeming social value on a three- to six-minute demo reel.

When some people do a tape, they attempt to superimpose little black "censored" boxes over the offending body parts or to blur or "mosaic" out the picture, like Reality TV, but this is more distracting than the nudity itself. Even if your scene is positively fraught with redeeming social content, it will end up looking vulgar.

If you make part of your living from films that include nudity, you will benefit from making two separate tapes and clearly marking the labels. If you've been working in what the Canadians once referred to as *Baby Blue Movies*—Playboy Channel fare—use very little of this work, and only the clothed segments. Don't try to edit out all of the nude shots, or you'll be editing for days and the dialogue will still sound like gibberish. It may be funny, but it's unlikely to get you work. Remember your audience. Many casting agents are female.

Typecasting

The film world works differently from the theater. In the theater, versatility is very important and highly prized. The more versatile your repertoire of characters, the more work you'll get. Once you hit Hollywood, it's the other way around. In the movies you must earn the right to play against "type." For those who feel their creativity is being stifled, think of it like this: Typecasting is a darn sight preferable to unemployment.

Too much versatility on your tape can be a problem because it makes it hard for casting people to get a grip on what it is you "really do." When you include a multitude of different characters, it's sometimes difficult for casting agents to peg you into a particular slot when decisions are being made. Keep in mind that they must choose between you and someone else, and if what you do gets a little hazy in their mind, they'll probably go with the performer whose tape is more specifically defined.

Casting decisions are often made on the basis of character types. Character actors regularly play a particular type. Most character actors have been around for awhile, and when a producer's looking for a character actor, they ask to see that type. When the breakdown service (a newsletter subscribed to by agents in which the casting departments announce their daily requirements), arrives in agents' offices, requests are made for those specific types. Agents respond by submitting clients who work in that category.

This doesn't mean that you should exclude different characters from your reel. It does mean that you should look for your best acting rather than the widest range when making your tape. If you do that and don't worry about being typed, the tape will have a better chance of getting you work because casting agents will have an easier time figuring out where you fit in.

There is an argument for actually typing yourself. An actor friend of ours came in some time ago to do his tape, and as we were sifting through the material, we began to notice a pattern. He was doing a number of one-day jobs around town at the time—a perfectly honorable way to make a living—and these different little scenes had something in common. He always seemed to be coming through a door and delivering some bad news. He often played a doctor in a hospital or another professional. Whoever he was playing that day, the news he brought was always bad.

In all of these scenes, he walks through the door with his head slightly bowed, sincere, concerned, learned, and calm, and says, "I'm sorry, Mrs. Whithers. We did all we could to save Johnny's sight . . ." or "I'm sorry, Carl. We lost her . . ." or "I'm sorry, Alec. The building permits for the orphanage were turned down. We didn't have a chance . . ." or "I'm sorry, all of you. There isn't going to be a retrial. . . ."

We decided to cut all those moments together into a single group and to tag the section with a little scene where he walks through a door, sits on couch, pauses, and says, "I'm sorry, Bill and Betty. I wish I could bring you some good news." It made an amusing segment of tape. The reel was memorable, and it made a point.

If you do manage to achieve a reputation for playing a certain type of character, at some point you will begin de-emphasizing that character and you and your agent will combine your marketing skills to break out of it. The agent can discuss your other qualities with the people who have been using you in a certain capacity and begin submitting you against that type.

You can do plays or showcases that demonstrate your range and of course you will revise your tape to emphasize the new direction. When doing this it is important not to lose your foothold in one area before you begin to establish yourself in another.

If you're still in the struggling-actor stage, why not work at getting yourself typed as something quite specific? Start getting known as a particular type to build up a track record, make some money, and show your folks you were right all along. Then you can start trying to break out of it. The theory, once again, is that it's a lot more fun to try to break out of typecasting than to try to break into show business.

Your tape is a sales tool used by your agent to interest casting directors in giving you work. Before you make your tape, discuss the scenes you intend to use with your agent, and listen to his input. Your agent knows what he saw in you when he signed you on, and if you make a tape that reflects those qualities, it will get you out more often.

If your agent is unfamiliar with your material, you may want to look at it together. Take the source scenes you have logged into the office, and go through them. If you have done your logging homework, you'll be able to fast forward from scene to scene, saving a lot of time. Counter numbers can be converted if necessary by measuring the difference in numbers between scenes. One hundred counter numbers equals six minutes, for example, so if your scene is twelve minutes into the tape, you will discover that's two hundred counter numbers on his machine. Logging your scenes at the outset will save hunting around later. You can then get through what he wants to see quickly and let him get back on the phone.

Your agent may tell you to use just the middle part of one scene or to put one scene first and another one second. You will save real dollars by going through this process first instead of making your tape only to have your agent say, "Why did you use that scene first? It should be last." If the latter occurs, you will have to return to the editing studio and pay to do it all over. If he wants your tape one way and you want it another, the easiest thing to do is to make two tapes. When you're in the studio making one tape, it's quick and inexpensive to make a second variation.

Preparing Your Source Material

Logging Your Scenes

By now you should have collected absolutely every bit of yourself on tape that you could conceivably use. It's time to log the tapes. You will need something to serve as a log sheet that will tell you what the show is, what tape it's on, how long it is, where it starts and where it ends, and any pertinent information. For example, you might note, "crying scene," "hospital scene," "good close-up for montage," or "too frumpy for beginning of tape." If you have lot of material, you will find brief but explicit notes of this type invaluable once you go into the edit bay and your mind is on a number of things at once, including the clock. You are welcome to copy the sample log sheet shown in Figure 4.1.

PROD. JOB # ______________

HYPOTHETICAL VIDEO PRODUCTIONS

Date ______________ Title ______________ Client ______________ Reel # ________ VTR # ________

SMPTE Code ☐ Footage Counter ☐

Scene	Take	Slate Information	Scene Starts				Scene Stops				Remarks	Circle Take
			Hrs.	Mins.	Sec.	Frame	Hrs.	Mins.	Sec.	Frame		

Figure 4.1. Sample Log Sheet

The better your preparation at home, where your time is free, the more time you will save in the edit session, where it isn't.

Logging Shooting Footage

For scenes, product demos, and video resumes that you shoot from scratch, you should log each take during the taping. It is a good idea at the time of the taping to slate each take with the scene number, the take number, and the counter number or timer number, which will be in hours, minutes, and seconds. This enables you to quickly find the tape you need during postproduction. Mark each tape clearly with the hour number as well as a description of the contents.

Numerous methods are used to keep track of scenes and takes when shooting on video. The most formal method—and the most dramatic—is to use the slate, or clapper. This is necessary when shooting on film and on some high-end video where sound is recorded separately from picture. To synchronize the two during postproduction, the exact frame in the picture where the little arm, which looks like a railroad crossing barrier, connects with the little black chalkboard is matched to the "clap" on the audio track.

The emotional power of these boards is irresistible. When working on some industrial shoots where we knew that the staff of the factory looked on us mainly as an intrusion, we pulled out the clapper and started slating shots, even though it was unnecessary for our time-coded video and our automatically synchronized sound. The minute the slate appeared, the romantic draw of that little board always turned resistance into enthusiastic cooperation.

We could write a chapter on the operation of a slate, the way scenes, pickups, and takes are numbered, and other related information, but here are the basics. First, scenes: Each new scene needs a new number. Each time you change the camera angle in the same scene, the scene number gets a new letter after it. Each time you reshoot part of a scene, add the initials *PU*, for *pickup*, after the scene number and letter.

Moving over to the "take" area, each succeeding take of the same scene, angle, or pickup needs a new number. If you are using time code, designate an area on the slate for the time code numbers (if there isn't one already), and write the tape time-code on the slate. The time code will be in either hours, minutes, and seconds or in simple counter numbers. Having written time code on your slate will be extremely convenient during logging.

In a third area on the slate, write in each roll, reel, or tape number. When using time code, remember to give each reel or load of whatever length an hour number and write that in. In the lower right-hand corner of the slate is a space labeled "Sound." If you are shooting picture only, with no audio, chalk in the letters *MOS*.

No discussion of the slate is complete without repeating the legend of MOS. A certain director of early talkies, no one can quite remember who, was an imperious and exacting fellow of German extraction. Because screaming was as much a part of his director's persona as his eccentric mode of dress, he would scream out, "Zis shot vill be mit out sound." The camera assistant could not resist, and on the slate, in three big letters he chalked: MOS. The joke spread to the next sound stage and the next. From there it traveled to all the other studios and out to filmmakers around the world. Today silent shots everywhere are always slated MOS.

There are many ways to slate without going to the expense of purchasing a clapper board. The person operating the recorder or camcorder on the shoot can use a log sheet to write down the time code or counter numbers of each scene and take. The operator will also write the designation *MOS* and any other special information where necessary. This could include *N/G* for *no good* or *E/S* for *end slate*. (You use the latter if you didn't have time to slate at the beginning or if you just forgot.) An end slate is always held upside down so the editor knows, without question, that it refers to the preceding shot.

A primitive but perfectly acceptable means of slating on video is to hold up your fingers in front of the camera. Do this once to indicate scene number, while saying it out loud, and a second time to indicate the take number. Later when you are logging the tapes, write these down on your log sheets and include your comments.

As you prepare your log sheets in advance of the edit session, note the preferred takes, their duration, and any comments you may have, such as "take 5 has best smile," "zoom starts too early" (it's not a good idea to cut on a zoom), and "talking in background." The more precisely you plan your finished product using the log sheets, the more time and money you'll save in the edit. Even though you're not working on an editing system at all at this point, what you're doing is part of the off-line editing process.

If your source material is on three-quarter-inch tape, you will obviously need a three-quarter-inch machine on which to play it as you log your scenes. Most performers do not have access to a machine of this type and will have to rent time at a studio or rent a machine from a video rental house to log their tapes. If you have a friend who owns one, be absolutely certain that it is in good condition before you use it. Tracking is particularly important. If the picture breaks up at the top or bottom of the screen and cannot be corrected with the tracking control, which adjusts tape position on the heads, or the screw control, which adjusts tape tension, remove the tape immediately. Do not use the machine until the tape transport, which guides the tape around the drum, has been repaired. This machine could cause edge damage on the source tapes and render them all unusable. Be particularly wary of the old wooden machines, and if there's any question in your mind, play an unimportant tape on it first as a test.

The most convenient way to log your three-quarter-inch tapes is to rent viewing time on a player only or playback only machine at a studio. They are cheaper to rent than machines that also record. The hourly rate should be quite low, and this might be a good opportunity to check out the studio. The studio may even give you a special deal if you plan to edit your tape there.

You can also rent a playback only machine from a video rental house by the day. They often give you two days for the price of one on the weekends, and if you have a friend who's doing a tape at the same time, you could share the cost. These establishments are listed in the Yellow Pages under "Video Equipment Rental."

Another alternative is to have the three-quarter-inch tapes transferred to VHS, which you can log at home.

Finally, the tapes you get from the studios on 24-hour loan can be transferred to three-quarter-inch tape if you can afford it, or you can book viewing time at the studio just before your edit session to log those scenes.

DEMO TIP: Know *exactly*—by counter number, tape timer, or time code—where each scene is, and know which tape it's on.

When you insert your source tape, rewind it to the very head, set the counter at zero, and go through the tape until you find each scene you're in. Even if you don't plan to use the scene, log its exact location. You may change your mind later, and a search during your edit session is frustrating, boring, and costly. If a tape contains SMPTE time code and the machine you're viewing the tape on can read the time code, you do not have to rewind the tapes each time and zero the counter. This saves lots of time, so if you have the option, use the time code. (Time code is a number that is recorded on each individual frame, and SMPTE is the acronym for the Society of Motion Picture and Television Engineers, who agreed upon the standard.) Time code is covered in Chapter 18. It is a tremendous timesaver.

DEMO TIP: All unlabeled tapes look alike.

Before you log your tapes, pick up a box of temporary labels in a stationery story, but not Post-it™ notes, which invariably come off. On the label, list every scene on the tape and its location in exactly the same way you listed it on the log sheet. Do this twice: on the tape and on its outer case.

Thinking about Results

As you endure the logging procedure, begin thinking about the actual design of the demo. This is not the time to make final decisions about what to use but instead to begin seeing your material from a different perspective. Ask yourself: What do I want the promotional video to do for me? The answer is: I want it to get me work. Every scene you view and every decision you make must answer two questions: "If I use this scene, will it get me work?" and "How can I place this scene in the body of the tape to give it the best chance of getting me work?"

If you are creating the demo tape on your own without a professional editing consultant, be as objective as you can. Try to see it through the eyes of an agent, a casting agent, or just a stranger. You will probably log the tapes on your own. If someone else is going through them with you, remember that this is *your* tape and it must make *you* happy. Listen to all opinions, then make your decisions on your own. We'll talk more about working with others on the design of your tape in subsequent chapters, especially in "The Third-Party Syndrome" in Chapter 6.

DEMO TIP: The happier you are with your tape, the more eager you will be to show it around. As a result, more people will see it sooner, getting you more work faster.

If you're excited about how great your tape looks, this excitement will rub off on your agent, who will then automatically do a better job of promoting you to the casting people. If, on the other hand, there is something about the tape that you're not that excited about or simply don't like, the opposite will occur.

You must consider your secondary objectives, too, and logging the tapes is a good place to start thinking about them. You may be in the enviable position of wanting to change the type of work you're getting. This is enviable because you are obviously getting enough of something. As you log your tapes, ask yourself the following:

- How can I use this material to get larger roles?
- How can I use this material to break out of soap operas and into nighttime television?
- I'm always cast as a cop. What can I do to change that?
- Even though I earned an M.D. before I chucked it all for Hollywood, I've only been cast as an airhead. What can I do?

- If they don't stop casting me as the resident psycho, I'm going to kill somebody! How can I get a role as a normal person?

Now is the time to start gently sifting through your work with the idea of finally emphasizing the insightful moment of the airhead, the compassionate or sane moments of the psycho, the on-location scene from the soap, the barbecue or family camping scene of the cop. For perhaps the first time in your career you have editorial control! It's a heady experience, so get carried away. The brevity of your final tape will work to your advantage.

DEMO TIP: The casting department considers three to six minutes to be the optimum length for a tape.

There is much more to designing your demo tape. Now that you've logged, labeled, and boxed your tapes, you're ready to begin.

5 The Fine Art of Demo Tape Design

Now that you have reviewed all your material and know where everything is by looking at your log sheet, you are ready to begin the creative process. If you have not yet chosen your editing studio, now is the time. The design of the tape will depend on the capabilities of the studio's machines, the experience of the editor, the hourly rates for the system you will be working on, and, most of all, your budget.

DEMO TIP: Know the charges for every machine, service, and add-on before you start.

When you begin calling around to various studios, ask them to mail you a copy of their rate cards. Every studio arranges its charges differently. This is not meant to confuse you; it's just that there's no set way of doing it. Pick out the services you are going to use, and compare them on each rate card. We've provided a sample rate card in Figure 5.1 to help you, but remember that the prices listed are not based on an existing studio. Nor are they suggestions as to what the actual charges should be. The prices are given to assist you in comparing the *relative* cost of systems and machines. Prices vary widely from city to city and from facility to facility. As stressed throughout this book, hourly rates are only one factor in the final cost. You must also consider time, the sophistication of the machines, and the experience of the operator. Prices go up and down as the engineering of the machinery evolves.

DEMO TIP: Your greatest price protection (after comparison shopping) is going into your session fully prepared.

Hypothetical Video Productions

Rate Card

EDITING:

On-Line Bay #1:

ASC Case controller, list management, 3/4″ SP, A/B roll, Grass Valley SEG, dissolves, wipes, fades, keys, digital video F/X, three-tube graphics camera, Chyron titles, For-A color correction, full-frame TBCs, 8-track audio, audio mixing,
with operator .... $145.00 per hr.
with additional VHS source deck .... $155.00 per hr.

On-Line Bay #2:

Convergence 195LM controller, 3/4″ or VHS A/B roll, Video Toaster F/X and titles, full-frame outboard TBCs with hue control, dissolves, wipes, fades, keys, three-tube graphics camera with operator .... $ 80.00 per hr.

Off-Line Bay #1:

Cuts-only 3/4″ or VHS to 3/4″ or S-VHS, Sony RM 440 controller, with operator .... $ 65.00 per hr.

Off-Line Bay #2:

JVC, S-VHS to S-VHS, ESC Super 90 controller with operator .... $ 50.00 per hr.
(A Panasonic controller and decks may be substituted.)

Cuts-only Sony Hi-8 to Hi-8, on-board title generator with operator .... $ 50.00 per hr.

DUPLICATION:

3/4″ or VHS to 3/4″ or VHS

Length in minutes:	1–5	6–30	31–60
1–10 copies	$6.00	$12.00	$20.00
11–20	$5.00	$10.00	$17.00
21–50	$4.00	$ 7.00	$12.00

Air checks, per half hr. .... $ 15.00

Prices do not include tapestock.

Figure 5.1. Sample Rate Card

Hypothetical Video Productions

Rate Card

STUDIO TAPING:

Includes lights and audio

Single three-chip camera, one-person crew ....$ 75.00 per hr.

Two ISO three-chip cameras, two-person crew ....$110.00 per hr.

Three-camera switched production four-person crew ....$160.00 per hr.

SINGLE-CAMERA ENG LOCATION TAPING:

S-VHS, Hi-8, or 3/4″ (four hr. minimum).......................$ 75.00 per hr.

Three-chip camera, Hi-8, S-VHS, or 3/4″ (four hr. minimum) $110.00 per hr.

Three-light lowel kit, add ....$ 10.00 per hr.

Additional crew members ....$ 50.00 per hr.

Travel time, beyond a thirty-mile radius ....$ 75.00 per hr.

A 20% discount is offered for all full (ten-hour) days booked in advance.

Each five-day week, booked in advance, will be charged as a four-day week.

TERMS:

- Prices subject to change without notice.
- Prices do not include tapestock.
- Charges begin at the prearranged call time, in studio or on location, and end when "wrap" is called.
- Overtime of one and one-half times the regular hourly rate will be charged after ten hours.
- Fifty percent advance deposit is required on all production.
- Twenty-four-hour cancellation notice is required for refund of deposit.
- All shooting tapes remain the property of Hypothetical Video Productions unless other arrangements are made.
- Shooting tapes and completed projects may be used in demos and promotions for Hypothetical Video Productions.
- Appropriate state and local taxes are charged on all finished programs.
- Hypothetical Video Productions' liability for loss or damage, however caused, is limited to replacement of raw tapestock.
- Payment in full of entire invoice is required before any masters, copies, or other tapes are permitted to leave the studio.

Call 1-800-HYP-VIDEO
Have a great shoot!

Figure 5.1. *Continued*

DEMO TIP: The most economical tape is the one that gets you the most work.

As you study the rate cards from various studios, compare the costs of cuts-only systems and A/B roll systems.

Cuts-Only Systems

In a cuts-only system, two tape machines are used to do the editing: a player (or playback machine) and a recorder (or record machine or record deck). These are connected by an edit controller, which is sometimes simply called an *editor*.

With this system, you will be able to cut from scene to scene and usually to fade up from black and down to black, but no other effects or transitions will be available. It may be possible to add a title or to shoot a head shot or a series of head shots with this system. This is the cheapest system available, and in many cases, it may be all that you need.

You might use a cuts-only system for the following reasons:

- You are on a very strict budget.
- You are making a tape for a specific purpose. For example, a casting director wants to see some tape of you playing a transient, and you have three such scenes on your collected works tape. You can assemble those scenes on a cuts-only system and send them over. This is one example of how handy a collected works tape can be.
- You are making a submaster for a specific purpose. For example, you need to send a tape to someone who doesn't want to see anything from a soap opera. You must edit around those scenes on your existing master, and a cuts-only system is perfect for this.
- You are making a standard commercial reel. This system will be perfectly adequate.
- You are a frequent talk show guest and are putting together a publicity reel of your interviews.

A businessperson making a video resume needs nothing more than a cuts-only system to do the job, although a title generator and a graphics camera—for adding photos or drawings of your projects—would be very beneficial. A graphics camera is simply any video camera used to shoot stills or artwork and is usually plugged directly into the system. The better the camera, the nicer the reproduction. For more information on cameras, see Chapter 14.

One major drawback of a cuts-only system is that color correction is not possible. If your source tapes are from a variety of places, they are likely to need color adjustment. Some will be on the red side, some a bit green, and so on. This is not too noticeable until you try to edit them together; then it becomes very distracting.

The Time Base Corrector

The professional solution is to run the tapes through a time base corrector so the hue can be adjusted and all flesh tones will look like flesh tones. A TBC is not always available on a cuts-only system, and when it is, the hourly rate is usually a bit higher for that system.

Some TBCs can produce freeze frames, which are also known as *still store* or *frame store* because the TBC stores the frozen frame. More elaborate TBCs are capable of digital video effects (DVEs). These allow for mosaic or tile effects and paint or posterization (which looks nice added to a final freeze), but they can also do much more. When you visit the editing house for your tour, ask your guide to explain the system's capabilities for these effects and the various prices that apply. Besides color adjustments, the TBC can improve brightness, contrast, and color intensity while correcting the technical anomalies that result from the less than perfect signal generated by videotape.

Some video equipment stores sell consumer special effects generators that produce similar effects to the ones mentioned above. At present, however, these effects degrade considerably on the copies, and even though they work fine for home video buffs, they are not up to the standards of professional studios. If you plan to use effects, find a studio with a true TBC.

A/B Roll Systems

The A/B roll system will boost the cost of editing your tape quite a bit, but if you can afford it, editing on this type of system is well worth the price, especially for putting together your edited master. An A/B roll system has two or more source decks or playback decks, allowing an A tape and a B tape to roll simultaneously. This makes possible a wide variety of effects (F/X), the most familiar of which is the dissolve. You can dissolve-in still photos, and the digital effects discussed earlier are on hand. A number of wipe patterns, which are used as transitions from scene to scene, are available, and you can add color backgrounds and split screens. The capabilities of this system seem almost infinite.

The most useful effects for your purposes are the ability to shoot stills, add titles, do dissolves, and freeze frames. The freeze frame has a myriad of wonderful uses. You can end your tape with a freeze of your best close-up and then superimpose your name while playing some flattering music underneath. Freeze frames are excellent in a montage sequence, and they can make a smooth exit from a scene that would otherwise cut abruptly.

The Character Generator

Character generator is the technical name for the title machine. It is often referred to in industry jargon as the *CG*. The character generator is an important consideration when creating your demo reel because nothing can make a tape look more unprofessional than the amateurish little letters generated by video typewriters.

One character generator, often referred to by brand name, is the Chyron. This machine is used to make the titles that appear on network television, but it is available in smaller, less elaborate versions for use in video studios. If the studio has a Chyron and its rental is reasonably priced, that's a definite plus. There are other very good CGs, though, so when you visit a studio, examine the titles on the studio demos.

Computers also have title-generating programs, and the brand most often used in your price range is the Amiga. If the studio has one of these in the system (and there's someone who knows how to use it), it's a good second best to the Chyron. The best software programs are Broadcast Titler, ProVideo, and the titling software included with Video Toaster by NewTek.

Video Toaster

Video Toaster deserves a thorough discussion here because it puts an enormous variety of broadcast-level effects in your price range. After years in development, Toasters finally began reaching studios in the spring of 1991. An Amiga with Video Toaster normally accepts four video sources, or inputs, such as VCRs, cameras, and videodisks. This means that as long as two TBCs are present, the program can do A/B rolls with over 130 wipe transitions. If your movie is out on videodisk, you can edit directly from the disk, giving you superb picture quality.

With Video Toaster, you can tumble the picture onto the screen or bounce the picture off. You can spin titles in or break them up into a million pieces of confetti. All of these moves will happen at your command. Effects Nirvana is a mere click of the mouse away. The title program is built in, and like all of Toaster's capabilities, it's user friendly. You can freeze or grab a frame, store it to disk and manipulate it infinitely.

Video Toaster can also serve as the edit controller if a special program

(Amilink by RGB Industries) is installed. This cuts down on studio space and makes Video Toaster an even more effective machine. Amilink comes with a complete list management program, which keeps track of edits for you like a log sheet but in a more sophisticated way.

Not all Toasters are ordered full-blown, however, and many are not capable of A/B roll. Some have not been equipped with TBCs, and that limits many functions. On-board TBCs are not capable of the color correction discussed earlier. When a Toaster is not fitted with a General Purpose Interface (GPI) cable, it cannot be triggered "from the board," that is, from the edit controller. This will result in some wasted time, as the operator will have to guess at the right moment to click the mouse, which could lead to several tries for each edit before it is correct.

When a studio has a Toaster, you're off to a good start. From there, ask about its capabilities so you'll know what you will be able to do if you choose to work in that particular facility. (Don't be hard on folks that couldn't afford all the options all at once. With the insatiable gadget appetites of videophiles, they'll probably be upgrading pretty soon!)

Format Choices for Your Master and Copies

Most studios still master on three-quarter-inch tape, and at present, this is the safest way to go. However, S-VHS is becoming very popular and will probably be the industrial standard quite soon. At the moment, you may still encounter some difficulties in getting S-VHS copied, whereas three-quarter-inch can always be done when you need it. Have your copies made on standard VHS so that they can be played on all machines. Videographers, production designers, and others who work behind the camera will need to make many of their copies on three-quarter-inch to retain the widest possible contrast range in the picture. Choose three-quarter-inch SP over standard three-quarter in order to retain the best picture quality. SP will playback in all three-quarter-inch machines. All production offices have three-quarter-inch playback machines, and this is where the tapes will most likely be viewed.

Types of Demo Reels and Demo Tapes

Now that you have an overview of the studio environment and what it has to offer, let's discuss the various types of reels you may be dealing with once you begin the creative process. First, a brief lesson in industry terminology. *Reel* and *tape* are synonymous. *Reel* is considered to be the more elegant term because it allows everyone to assume that you hobnob with the gods of the movie world instead of the ordinary workers in video. *Reel* helps us feel like we're in a profession with a history, which, of course, we are.

Sample reel. This is the most elegant synonym of all for *demo tape*. It has always been used by directors, lighting people, and camera operators. We heard it first when Kent Bateman called up and said, "I want to come over and add some scenes to my sample reel." We quickly realized that he meant his demo tape. There probably wasn't even a pause in the conversation. A *director's reel* is the same as a *sample reel*, which is the same as *demo reel*, which is the same as a *demo tape*, which is the same as a *presentation tape*, and so on.

Compilation reel (also known as *collected works tape*). To create a collected works tape, edit *all* your scenes from your air check tapes. Keep everything you may ever want to use at any time in the future, even if you can't see a use for it right now. Don't bother to do the refined internal editing you will do when you create your edited master. Lose the big sections you're not in, and let it go at that.

Once you have everything on one or two tapes and you have made sure that the picture quality is okay, which it should be if the off-air recording was done properly, you can recycle the source tapes you know you'll never watch again. This saves time when you come to edit your demo, which in turn saves money. You can reuse the source tapes and run far less risk of misplacing important tapes when you finally do your edit. If you've ever lived in a Hollywood studio apartment, you know how quickly those big three-quarter-inch tapes can pile up. Be sure to save important, valuable, or old source tapes, but those ten episodes with six lines each on "The Young and the Restless" can certainly go.

It's a wise idea to make a protection copy of your collected works on three-quarter-inch. If you want to be really safe, copy it to broadcast Betacam tape, a very high quality tape used in broadcast studios. Store this tape in the safest possible place you can imagine, like a bank vault or the family farm back in Nebraska. Then, if something happens to your life's work stored on the original, you have a backup.

Calling card tape. A calling card tape is a very brief tape of one or two minutes. It often consists only of a montage or of one or two very impressive or well-known scenes. It is sometimes easier for agents to get casting people to look at one of these, especially if they've never seen the actor before. Calling card tapes are generally sent out as an introduction without a specific role in mind. Some people make these tapes if they are planning a saturation campaign.

Industrial reel. Quite a new phenomenon is to keep your industrial work separate from your broadcast television and feature film work. When your agent gets an industrial call, he or she will send out the industrial reel. There is no harm in adding mainstream work to an industrial reel. The better known you are, the more the corporate folks will want you to plug them.

Single-purpose tape. If a casting agent wants to look at scenes of you as an auctioneer, assemble these scenes from your demo master or collected works, keeping the editing simple, and send over a copy. If you don't have any auctioneer scenes on your tape, consider shooting an audition scene. Check with your agent. Don't forget to label the new tape and to rewind and label the copy.

Special skills tape. If you have a special skill—for example, if you are a black belt in Karate—rather than put nonbroadcast footage showing off your skill on your main demo reel, make a separate special skills tape. Send it out with your regular demo when being considered for roles that require your skill.

Commercial reel. A commercial reel can be very simple, with your best, latest, or most memorable commercials at the beginning. It is acceptable to put entire commercials on the tape, fading up and down between each one. Be sure to watch the color carefully, and leave off any with low production values.

You may wish to edit your commercials so that they are shorter and feature you more. You can lose the end card that plugs the product with the announcer's voice-over. Keep in mind that people who make commercials are not offended by plugs, so don't work too hard to eliminate them if the result will look choppy or confusing.

Vanity tape. Home movies and video, personal snapshots, church outings, the pet rabbit, and photo montages of your wedding—however dreamily produced—are strictly for family and friends.

Celebrity reel. A wide range of professionals require celebrity or publicity reels, including the following:

- authors being considered for talk shows, book promotions, and tours
- "media psychiatrists" who appear on news shows, talk shows, educational shows, and video magazines
- actors who are required to promote their television shows and movies
- trainers and handlers who make appearances with rare or interesting animals
- scientists
- politicians
- environmental activists

These are only a few of the celebrities whose activities are advanced by television appearances.

6 The Creative Process of Demo Tape Design

As in any creative process, once the rules of making a demo tape are understood, it is permissible—perhaps necessary, sometimes even elegant—to break them. However, the aim of this book is to show you why we do what we do and how we do it. What you choose to do after you understand the rules is up to you.

Choosing an Editor

DEMO TIP: Choose your editor as you would choose your photographer.

When you look at a studio's sample demo tapes, find out who edited them and whether you will be working with that person. Does it seem like you and the editor will get along? Do you think the editor's style will work with your material? Has the editor been recommended by anyone? Does he or she seem to have a command of the machines and the craft? Is the editor excited about the work you are shown? Does he or she respond openly to your comments?

Editing your tape can and should be an exciting, pleasurable, and rewarding experience. The more comfortable you feel in the editing facility and the more at east the people you will be working with appear, the more likely your experience will be rewarding and your tape the best it can be.

It is hoped that you will find an open and friendly soul with a high degree of technical ability, a superb eye for movement, a positive attitude, a love of theater, a complete disregard for their own physical needs, a lightning quick mastery of the equipment, the capacity to make creative suggestions without appearing to be critical, an astute understanding of what the tape is intended to achieve, and a personal mission to keep you on time and under budget. But, just in case, let's assume the person you're up against has scrawled three large words on the wall in front of the edit bay: "You tell me." For the purposes of the next two chapters, we will assume that the editor or operator has no creative input in the design of your demo tape.

Opening Your Tape

Let's begin from the very head of your reel with the leader. The leader is considered to be so important that major agencies spell out their requirements to us in precise detail and brook no deviation. When they go in with a tape, they are about to make a sales presentation to the casting people of a very expensive product (you). Everything possible must be under their control. If they have finished saying everything they have to say and there is still no picture to look at on the monitor, it becomes embarrassing for everyone. After discussing the leader, we will move on to the placement of the agency name and your head shot.

When viewing your tape, the first clip prospective agents and casting people see should be your best, most exciting scene. Alternatively, you can construct a short montage. Like a teaser, a montage is a series of quick cuts, often set to music, and used to get across a variety of ideas in a brief period of time. The opening of your demo should interest viewers in your work and entice them to see the entire body of the tape. Montages have worked very well for our clients.

The Leader

Some editors are determined to put color bars on a tape, and they have good technical reasons for doing so. However, casting people hate color bars because they've had to sit through a minute or more of them too many times. Staring at nothing for a whole minute can seem like a very long time when the producer is breathing down your neck. There are three reasons to omit the color bars. First, even if you lay down only five seconds of bars, the casting people won't know that, and you'll immediately be off to a bad start. Worse, as soon as the bars appear, someone might hit the scan button and go forward into the body of your tape, stopping who knows where? There goes the impact of the montage or that brilliant opening you worked so hard to achieve. Finally, the only purpose color bars serve—unless you're in an editing suite—is to adjust the color on the television set. In a casting session, no one ever does that.

There is one exception that should be considered. You may want to lay down bars on your edited master. This will aid TBC adjustment in the future when you use a submaster to update your tape. The editor will forward the tape anywhere from thirty seconds to a minute before starting the actual program. This is partly because the highest dropout level exists at the head of the tape. Dropouts look like little pinhead-sized shooting stars that flash across the screen, and they are caused when minute particles of oxide come off the tape.

When it's time to order copies, you can instruct the person who runs your dubs to start the copies after the color bars. Beware, though. There is a strong possibility that the person who takes your order is not the one who will run your dubs, and duplication facilities can get very busy. Remember, the more hands that your tape must pass through, the more likely it is that a mix-up will occur.

Unless you are a video professional, skip the color bars and put up with a few dropouts. Your material should begin after six to ten seconds of black leader.

Your Agent's Name and Logo

DEMO TIP: Include your agent's name or logo at the beginning or end of your tape. Ask first to be sure your agent likes the idea.

If you include your agent's name at the beginning or end of your demo, your tape may be sent out more often. The agent's name is usually placed just before the head shot and allowed to remain for about two seconds with no audio. Alternatively, you can bring it up at the end. There is no harm in including the logo at both locations.

Some management firms insist that their logo be included on tapes they send out. Your agent might be able to provide black-and-white artwork that you can take to your editing session. It will be recorded, or keyed, onto the tape when the editor shoots your head shot. Otherwise, a logo can be keyed from a letterhead or a business card. If the colors of the logo do not have enough contrast to key well, photocopy them in black and white, and shoot the copy. If you change agents or managers, it will be a simple matter to replace the logo.

The Head Shot

DEMO TIP: Begin your tape with the head shot you send out most often, and superimpose your name over it.

Why begin your demo tape with a head shot? Consider the following scenario: The second your video appears on the screen, the casting person, the producer, or another important person immediately says "Who am I supposed to be looking at?" A discussion ensues while your best scene is being shown. At the same time, the viewers are also trying to find a spot to put their coffee cups down and are getting ready to make some notes. If other tapes are being played in the same session, conversations take place while one demo is pulled and the next is inserted. No one looks at the screen until something happens to draw their attention back.

By opening with a head shot, you give your viewers time to settle down and show them who they are "supposed to be looking at." Bring the head shot up for a maximum of three or four seconds, and you can run a little music underneath to help draw the viewers' attention to the screen. As a result, someone will be watching when your opening scene or montage comes up.

It is not a good idea to put your phone number on the tape. It tends to make your demo reel look like a late-night television commercial. Instead, type the phone number on the label, and leave it at that. If the number is on the agency's artwork, however, by all means include it.

If you decide to change your head shot later on, it will be easy to do. Simply have the editor insert the new head shot in place of the old one. This takes very little time, but check on the charges beforehand. They will probably be reasonable.

The Montage

A montage is a brief series of cuts that are edited together to form visually intriguing glimpses of the performer. It is intended to interest viewers enough to pull them into the body of the tape. For more information on montages, see Chapter 7.

DEMO TIP: Do the montage last.

Put together the body of your tape first, then create the montage. The reasons for creating the montage last are both practical and psychological. Practically speaking, this gives you and your editor time to get more familiar with your material and to start working out ideas for the montage while getting other things accomplished. It saves an enormous amount of time on the clock.

From a psychological point of view, a decent montage takes from two to three hours to create, and if you and your new editor have been working for over two hours and haven't finished the first thirty seconds of your tape, you might panic and your editor is going to be either embarrassed or defensive. Both of these feelings are a block to creative endeavors. You might ask yourself, "If thirty seconds takes this long, how long is six minutes going to take?" The body of your tape—the following five and a half minutes of the reel—will be an unknown, and you will be understandably nervous about how everything's going to look. If you add all this to the stress of reviewing your life's work in front of a stranger, it's obvious that the edit bay can turn into an emotional pressure cooker. The truth is the body of the tape is a cinch compared to the montage, so leave your montage until the end, and you can always cut it short if you are running overtime. The whole process should be fun and artistically rewarding. Doing the montage should be dessert.

If you decide to do a montage, leave some space on the tape, and come back to it after you have created the body of the tape. Thirty seconds is just right; forty seconds should be the maximum.

Your editor, who may have some experience in doing montages, can make suggestions as you go along concerning what would cut together well. You will save more time by asking the editor to read out the time code numbers where each possible shot can be found. Log the numbers and jot down the type of shot so it's easy to find later. If the edit controller has an electronic notepad, your editor may be able to do this for you, but remember that editors do have a lot of other things to think about. It might be better to take on this job yourself.

No matter how detailed a plan you make in advance, be prepared for inspiration when you go back to do the montage. You will restrict your creativity and effectiveness by being too rigid about what you decide to use.

Selecting a Theme and Enhancing Style with Transitions

DEMO TIP: When making your reel, try to evolve a theme and a style.

Theme is the chosen subject of a creation and can be repeated throughout the work in variations. A style is the manner in which something is done or the shape or design of a creation.

When making a tape for an actor whose forte is portraying villains, we eliminated any shot or scene that didn't suggest *evil*. We kept that word in mind while we worked on his tape. That was the theme. For another client, an especially endearing young actor, we chose *warmth* as a guide for her scenes. While presenting the broadest range possible, we measured our choices against that theme and deliberately picked scenes that contained that feeling.

In the example that follows, our theme is the presentation the actor distilled to the essence of his artistic honesty. Whatever character he is portraying, the inner truth of that character unquestionably emerges. We went for the peak moment of each performance and set ourselves the difficult task of using only that moment. The style we chose, or manner of presentation, was one of simplicity, brevity, and pace.

A few years ago, a brilliant character actor came to us with a hopelessly long demo tape. It ran for an entire hour. He had begun to realize that he was only being cast as the first few character types on his reel. We mulled over the tape for several days without making an edit. It contained scene after scene of excellent work, and there just wasn't anything we felt we wanted to cut. He returned to the studio, and we began to work on it together.

We decided on a reel that took only the essence of each role or situation and never repeated a character, a style, or a feeling. We reduced one scene after another to only one line. His acting was so rich and believable that the single line sometimes conveyed the entire meaning of the scene and the depth of the character he had created. The tape was structured like a montage, but we added no music or titles or anything else other than a title card with his name at the head and a freeze frame with his name at the end. To help people recognize the movies represented on the reel, we included a reaction shot, a line, or perhaps a quick two-shot with the many stars he had worked with. The tape had such a pace and was so intense that it created an excitement all its own. When viewers recognized a famous movie, they yelled out its name. By the time the tape was over, everyone was left with a sense of the versatility of this actor. We finished with a tape that did the whole job in one minute and forty-eight seconds.

DEMO TIP: The more work you have, the less elaborate the style of your tape should be.

The style of your demo reel refers to how you put together your scenes, which can enhance or detract from the image you're trying to convey. You reflect your style by such things as the transitions you use, how you open and close the tape, and the pace you maintain between scenes. If you successfully use transitions to support your style, you will project the image you've planned. Dissolves slow the pace and give tapes a lyrical quality, while straight cuts pick up the pace and develop excitement. Wipe transitions can do either, depending on the speed, or rate, of the effect, but they can also be distracting.

Bridging cuts with audio. Audio can be used effectively to bridge cuts in the middle or at the end of scenes. Cross fading the audio from one scene to the other can be a simple matter: Just bring the new audio up before the cut, and fade the old audio out just after. It's also possible simply to bring the new audio in just before the new scene comes in. This gets viewers ready for the new environment they're about to experience. For example, you are saying goodbye to your lover at a quiet little park in upstate New York. It's early evening, there are birds chirping, crickets, a rustle of wind. As you slowly part, holding each other's hands until the very last moment, the noise of the A train creeps slowly up on the audio until it's finally a screeching roar and we cut to you squished in a crowd of passengers on your way to dance class Monday morning. This technique is used extensively in film and television. You might try watching for it the next time you're at the movies.

Transitions between clips from different sources can be accomplished in a number of ways. You can

- use the show titles directly off the air check
- add your credit if it comes at the top
- superimpose the show title, using the character generator, over the first shot of each scene
- insert the show title between the scenes over a colored or black background or by using the CG
- make a title and superimpose it over an appropriate freeze frame from each scene
- use no titles and simply fade in and out from one scene to the next
- use no titles or fades and cut the clips together, one after another, without stopping
- generate a title card at the head of the reel, listing the show titles in order and the duration time of each clip (directors, producers, and cinematographers favor this one)

Some types of transitions should be avoided. Introductions shot in the studio tend to clash with real television scenes, and no matter how well they're done, they end up looking amateurish. If your entire tape is nonbroadcast video, a taped intro is okay, but keep it very short, no more than a couple of sentences. Do the intro straight into the camera, with a nice smile, then fade out and come up on the scene. If you do that, it's permissible to come back on camera at the end, smile, and say, "Thanks for watching."

Always keep the following in mind when selecting your transition style:

- Jokes, gags, and outtakes, including remarks on the audio track, are generally objectionable to casting agents.
- Interludes between scenes, other than a title card, will probably get your tape pulled.
- Once they've heard you speak, attempting to insert a montage in the

middle of the tape usually feels like down time to casting. Once they're looking at your acting, keep them looking at it.

- Never introduce your tape by sitting in a chair in a studio and reading your resume.

We have seen some exceptions to the above that were amusing and successful, but they are very few. If you decide to risk it, consider who will be watching the tape and under what conditions. Remember that if you come off the least bit flaky, someone will yell, "Next!" For more information about on-camera performance, see Chapter 12.

DEMO TIP: The less you have to put on your reel, the more you can use titles and special effects to enhance your professional image.

Don't overpower yourself with dazzling special effects, however, and it's most unwise to make up a title that says you starred in a film if you didn't. Luck will have it that the first person who sees your tape will be the agent who cast the real star of the show, and there goes a day player job that might otherwise have been yours.

That said, there is no harm in including your real credit if it appeared at the top of the show. A neat way to achieve this is to run a little of the music from the show, say four seconds, then do a video-only insert of the title card for two seconds. Cut to your card for two seconds, fade down, then up into the first scene. It's fine to choose a new background and remake the credit, but don't do this with a cheap title machine or people will think you're trying to fool them, even if the credit is true. On television, a guest star credit is often superimposed over some action by the series regular. This does neither party justice and occurs only because that's how the credits came up on the Chyron. It's perfectly acceptable to choose a background scene that features you or to freeze a shot you like and remake the credit so that it appears over your own work. Every tenth of a second on your tape should work just for you.

One good reason to use your billing if it occurs at the top of the show is that it helps your agent negotiate that credit again or do better. The casting people will know you're in that category when they see it on your tape. End credits are of no use, except in the video you make to send your folks at Christmas. Everyone gets an end credit, and you do not enhance your professional image by putting it on your tape. Shared cards aren't much better because your credit appears with others which doesn't improve your stature at negotiating time.

Structuring the Body of Your Tape

Length

Remember the demo tip that says that casting considers the optimal length of a demo tape to be three to six minutes. It is generally believed that the attention span of a captive audience watching a promotional reel for a product or a company is six minutes. All the information you want to present should be conveyed within that time frame. This rule also applies to casting agents and others who must watch demos of all types.

When you logged your material, you used either the timer on your VCR or a stopwatch to determine the length of each scene. If after timing every scene, you still can't get everything below eight minutes, don't worry. In the editing suite there will be myriad ways to shave those precious seconds.

DEMO TIP: The tighter your tape is, the more impact it will have.

When you structure a tape, your objective is to get the casting people to watch the whole tape through, heave a big sigh at the end, look at each other, and say "Whew! That was some show!" without even thinking about the length. If you work at getting your scenes cut from eight minutes to six, or even less, you will be going a long way toward that end.

DEMO TIP: Edit the scene to its essence.

To achieve your objective, you'll need to edit each scene to its essence. If you have a four-minute scene, even if it's all pretty good, find the very best forty seconds or a minute, and move on. You can do this by using one complete segment or by cutting the scene internally until you've eliminated everything but the very best of your performance. Once viewers feel that they know what the scene is about and what you're doing in it, they are ready to see something else. They do pop the tape if nothing new is happening.

Selecting Scenes

All scenes you choose should provide the following information about you:

- You can act (another way of saying you have talent).
- You know your lines.

- You are at ease and comfortable.
- You have a presence.
- You can command a scene.
- You are at ease with the star (unless you are the star).
- Your performance is as real as it can be.

Let's assume you're someone with a bit of everything from which to choose. There might be classroom scenes, nighttime television, a video of a stage play, some soap opera roles, a movie role or two, and some commercials.

Classroom scenes. Avoid classroom scenes if you can, and eliminate them as soon as you have enough real television or film. When using the class scenes you already have, select them carefully. Look for the best possible audio, color, and loose close-ups, choosing segments that are mostly you. Pick the best thirty seconds to one minute that you have. If you seem to be overacting, discard the scene and shoot it again. The same goes for bad audio, hissing or muffled dialogue, bad or unflattering lighting, a continuous extreme long shot, and in most cases a continuous extreme close-up.

One renowned acting coach in Hollywood uses the extreme close-up for instructional purposes. If you decide to put the well-taped material from these classes on your reel, use only your very best acting and make the segment brief to keep the shot itself from getting boring. If it's important to you to use several clips, intersperse them with other material.

Stage plays. When using videotape of a stage play, the same considerations apply. Even if the play was shot by PBS, it will have a very different look and quality from television or film. Acting styles and levels are automatically different when you're reaching out to a big house as opposed to drawing the camera in, and they can appear to a casting director as overacting. Overacting is one of the biggest fears filmmakers have, and casting directors just won't take the chance. A director can do something with underacting, an editor can do something with nonacting, but no one can do anything with overacting except leave it lying on the floor.

Industrial work. Industrials may be used with care, and the level of production value will determine their usefulness. At one time they were used only as a last resort, but their quality has improved so much recently that attitudes are changing to the extent, in fact, that agents have begun asking for a separate industrial reel. When choosing scenes from your industrial work, avoid silly jokes intended for the amusement of the company's employees, and look for writing that gives your acting a chance.

Soap opera scenes. Soap scenes present an interesting set of problems. Because of the time constraints and set limitations of soaps, they are necessarily much more static than nighttime television. This gives soap opera scenes a sameness even when the content of the story is completely different.

Another intriguing factor is that the characters always seem to be lying to each other. For example, an actor doing her tape with us will often say, "I want to use that scene because I'm being deceitful to Johnny in it." Unfortunately, for her performance to make sense, you need another scene, which she's not in. Her scene might go like this:

> SHELLEY: Oh, Johnny, I know how terrible this all looks, what with nice old Mr. Simmons in a coma and all. . . .
>
> JOHNNY: It happened in the factory you inherited from your Dad.
>
> SHELLEY (very sincerely, touching his arm, and moving closer): I know, but can't you see they just want you to think I'm involved? They're trying to split us apart, that's all. You can believe me, Johnny.

Without the scene in the hospital in which the old man comes to just long enough to whisper that Shelley was standing right there when he was pushed into the vat, how will anyone know she's lying unless she winks at the camera?

Choose strongly contrasting scenes from your soap work, and assume that whatever the content of the scene you select, that's all there is to the story. Use an angry scene, then a love scene, or a happy scene, then a sad scene, and so on. If you have a choice of scenes that are equal in other respects, pick ones that were shot on different sets. Remember that the soap "look" will work against you.

Unless soap opera work is all you have, it's best not to open your tape with it. Two scenes cut in somewhere down the line is usually plenty to show what age you play and a bit about your range. One exception, of course, is when you make a tape specifically for a soap opera casting department. In this case, you may want a reel made up entirely of soap opera scenes. You could add some nighttime television and feature film footage to this reel if you have it.

Other broadcast sources. For multicamera sitcoms, nighttime television serials, movies of the week, reality television dramatizations, and film, you have much more latitude. Make sure the clips are varied, and don't repeat yourself. Once the casting agents think they've seen all you can do, they'll pop the tape. If you play a range of character types, don't do the same character twice. For example, if you play a bad guy sitting in a chair menacing someone with a gun, don't follow it with another bad guy menacing someone else with a gun. Use the sitcom next where you're delivering flowers and you get invited in. Then add the movies of the week where you are the lab technician who discovers the virus. Keep the tape moving; always add something new to keep viewers interested.

For series regulars and stars, use some of the most memorable moments of the show, particularly those that got a lot of response from the industry or the public. A montage with dialogue and, if possible, the theme music from the show mixed together can work very well. This montage might be useful when doing guest spots on talk shows. You may decide to put together an entire tape made up only of your work on the show. This is fine, but it's also beneficial to have a general reel as well that also includes material from the show.

Scene Order

DEMO TIP: Put your very best scene first.

DEMO TIP: Put your second best scene second.

When structuring your tape, your objective is to keep viewers watching until you want them to stop. Casting people will pull the tape the second they have made a decision. For this reason, you must put your best scenes at the beginning of the tape. Sure, you would like the casting people to sit through the whole tape—they'll miss some great acting moments if they don't—but if they pop the tape in order to give you a job, what the hey? If you save a great clip for the end of the tape, it may never be seen. Hit them hard with your best stuff, and hit them all at once.

Remembering that your best scene should go first and your second best second, having organized your tape for the variety and impact mentioned earlier, and all other things being equal, the general order of your source material should be as follows:

- feature films
- MOWs
- nighttime television (series, reality show, or sitcom)
- soap opera
- industrial
- nonbroadcast video

Commercials can be interspersed with the above in a lighthearted or witty way. Any humor you use on the tape should be aimed at a general and conservative audience. Remove all the messages and product plugs that don't include you, and cut the final plug at the end by simply fading to black.

Editing Scenes Internally

It is especially difficult to know what to leave out when editing a scene internally. Remember, the goal is to feature your best work, not to show every little bit of work you've ever done. Some of the shots that you don't want to include on your tape are the following:

- fight scenes, especially if they feature a stunt double, unless you want to be the one thrown off the building in flames next time
- lengthy love scenes, especially if there's a body double, unless you want that type of work
- all but your cue lines from other people's dialogue
- other people's close-ups
- coming in doors and going out doors unless it's an honest to goodness entrance or exit
- unrelated cutaways such as the coffee starting to drip
- establishing shots
- conversations that don't include you
- moody camera pans of the scenery
- long walk ups, waiting shots, and wide shots where no one can tell it's you

---

DEMO TIP: (Almost) no one should be on your tape but you.

When editing a scene internally, remember that *you* must be the performer featured on your tape. To favorably impress the people who are in immediate control of your destiny, don't make them grouchy by forcing them to watch someone who isn't under consideration. If they have to wait too long to see you, the tape could come out. Not only that, but if you're sharing a scene with someone else of your type, the last thing you want said is, "What do you think of the other one?" Finally, including other people on your tape dissipates the intensity of your performance, and what the casting agents should be looking at is your performance—nothing else.

DEMO TIP: Do not attempt to tell the story.

Your tape is your audition. Keep the structure of auditions firmly in mind. Remember how the reader reads the opposite lines in a live audition? They often read just the first and last line of the other character's speeches. The casting agent, the producer, the director and maybe even the studio accountant are staring only at you. It's possible to lose sight of this when you edit your tape.

Jump cuts. You can see right away that you're going to have some problems cutting all of this together. The first problem you will encounter is the jump cut. Let's say you're looking right to left and then suddenly you're looking left to right. That's a jump cut.

Suppose that in your scene you're sitting at the table talking to "Mom" and she gets up and moves to the sink. When you edit out her walk, she disappears from the room like magic. This distracts from your acting, and even if the jump is not too noticeable, it can make viewers feel uneasy without revealing the real problem.

There are several ways to solve the problem of jump cuts. You can

- add a reaction shot, perhaps yours or a brief moment of someone else's close-up
- go to a quick cut-away of the coffee dripping
- leave part of Mom's walk but cut most of it, pasting it together by inserting a close-up of you
- drop the earlier part of the scene where you're looking to the left
- drop the latter part of the scene where you're looking to the right
- drop the middle part of the scene where Mom's at the sink, and come back in once she's back at the table

Chronology. When you edit a scene for your tape, it need not be put together in the same order as in the show. We edited a scene for a fine actor whose challenge was to be the catalyst for Albert Brooks's tirade in the brilliant, "You can't fire me—I quit" scene from *Lost in America*. Tom Tarpey was superb as the vice-president from New York who sits on the couch and joins in the conversation every few minutes. For his demo tape, the order of his lines was entirely irrelevant, and we shifted everything around so the words made sense and no emotional moments were out of place. By switching his lines around, we were able to reduce the scene from seven minutes to a single minute. For the purposes of this demo tape, the scene made perfect sense. In fact, even we couldn't reconstruct many of the changes without going back and looking at the original. There were no jump cuts, and the editing was seamless. Viewers of Tom's tape learned all they needed to know, and it saved everyone a lot of time.

Salvaging a scene. Using the editing tools described above, take a look at the scenes that you wanted to use but decided had problems. Maybe there were some jarring lines in an otherwise usable scene. Consider the following example:

Fade in.

YOU (emotions building): Mom, I really wish you and Dad would try to be more understanding when I bring my friends home. Last night he acted terrible just because Jerry has pink spikes in his hair. It's just not faiyyyyr! SCREAM, SCREAM. TIRADE, TIRADE, TIRADE. I mean, it isn't like he's on drugs or something!

MOM: I know, dear. Your father's very conservative.

YOU: You've got to taaaalk to him!

MOM: I will, dear. Now, hurry up. You're going to be late.

Fade out.

The problem in your scene is the scream scream scream, tirade tirade tirade part. If that section were gone, you could use it. Let's try this:

Fade in.

YOU (emotions building): Mom, I really wish you and Dad would try to be more understanding when I bring my friends home. Last night he acted terrible just because Jerry has pink spikes in his hair. It's just not faiyyyyr!

MOM: I know, dear.

YOU: I mean, it isn't like he's on drugs or something!

MOM: Your father's very conservative.

YOU: You've got to taaaalk to him!

MOM: I will, dear. Now, hurry up. You're going to be late.

Fade out.

Notice how we switched Mom's "I know, dear" to solve the jump cut? Here's another solution:

Fade in.

YOU (emotions building): Mom, I really wish you and Dad would try to be more understanding when I bring my friends home. Last night he acted terrible just because Jerry has pink spikes in his hair. It's just not faiyyyyr!

MOM: Your father's very conservative.

YOU: I mean, it isn't like he's on drugs or something!

MOM: I know, dear.

YOU: You've got to taaaalk to him!

MOM: I will, dear.

Fade out.

Reversing the "conservative" line may work better once your offending tirade is gone. Try it one way, then the other. Either way, the scene is cleaned up without losing continuity. Notice that we dropped Mom's "Now, hurry up. You're going to be late" line and saved a second, tightening the tape.

DEMO TIP: If there's any doubt at all in your mind, drop the scene.

Making a judgment call. If you cannot fix a scene to the point that you're completely happy with it, leave it off the tape. It will subdue your enthusiasm for showing your tape, no matter how hard you try to mask it. You'll be tempted to apologize for the scene, which is distracting and suggests you lack self-confidence. If it's the only scene you have or if there's some other reason why it has to be included, use it, but replace it as soon as you can. There is some possibility that you are the only one who doesn't like the scene. The others may be right, but if it's bugging *you*, that's what counts.

Remember the Hollywood starlets of the thirties who would only allow themselves to be photographed from their best side? Everyone has similar phobias to one degree or another and it's important not to let them get in your way. If you don't like a scene because you are self-conscious about some part of your anatomy, that problem should be addressed and solved. The cheapest and simplest solution is to realize that you're the only one who can see it and to learn to forget about it. If the feature is truly noticeable, you could try generating a little Jimmy Durante pride in it!

Making Realistic Creative Choices

DEMO TIP: Some things are worth the time and the trouble. Some things are not.

One media celebrity had a problem with her early interviews. She realized she had a habit of saying "um" all the time. We didn't find it that distracting, but she hated it. She wanted us to go through her tape and edit out all the "ums." This was a job that could have taken days and resulted in extremely distracting jump cuts or the prospect of lips moving without sound if we tried cutting out the audio only. The solution was to forget trying to use the whole interview, to use instead the sections with the fewest "ums," to send her to an "um" therapist, and to wait until she taped another interview.

The sun never shines in an edit bay, and perhaps that's why it's easy to lose perspective and waste time on things that don't matter. You can usually find a simple way out of a dilemma if you think about the problem objectively. Instead of wasting hours on a scene that was shot mostly over your shoulder on the other actor, necessitating a hundred and fifty edits to create a scene that's of any value, use a shot or two in the montage and forget the rest.

DEMO TIP: Limit your creative choices.

If you decide you want a different effect for every transition—a wipe, a star, a million little squares, a push-off, a pull-on, and who knows what else—you could spend hours choosing exactly the right effect for each scene. This is going to drive the price of the tape through the roof. Instead, pick one effect or two or three, and use them throughout. But leave it at that. Unless you're cutting the Super Bowl, no one will care, and you will save big bucks.

The same philosophy applies to the Chyron. You can get involved in picking just the right color for each title—your choices could be as high as sixteen million—and before you know it, you've lost all sense of time playing around with something that's taken on astronomical importance. Forget it. Look at a few colors, pick one that looks nice and reads well, and get on with what you came to do (editing your demo tape). Again, money will stay in your pocket.

DEMO TIP: Choose title colors that will contrast with the scene, and always use a border.

Make the titles sharp and legible by choosing colors that contrast with the scene. Unless you use a white border around them, avoid using reds and purples. Those colors will bleed on the copies, and the words will be hard to read.

Remember, this is a demo tape of your work in your profession. All creative choices should enhance your work, not upstage it. Let your editor make a *pièce de résistance* with his or her own demo tape.

DEMO TIP: Keep control of the creation of your tape.

Some video houses allow you to drop off your tape for editing, and they will call you when the reel is done. If you have dealt with the establishment for years and your rapport with the studio is excellent, your tape will likely turn out fine. However, if there is any question at all in your mind about seeing eye to eye with the studio, be there to make sure that your requirements are met. Once the tape is finished, it's your responsibility to pay for it, and if you need to change it, you will have to pay again.

You may want someone else to handle the job because you believe you are not a good judge of your own work. Many actors feel this way, and it's understandable, but at the very least, you know what you don't want. It's important to you and to the operator to make sure your most specific views are known.

You want to be so proud of your tape that you can't wait to circulate it. Again, if there is even one close-up that you hate, you will feel reluctant when you're asked to show your tape.

Of course, retaining control of the creative process does not mean refusing to accept input. It simply means that your tape should reflect you at your best the first time you edit it, and the best way to achieve that result is to stay with it.

DEMO TIP: Cutting your tape twice doubles the price.

The Third-Party Syndrome

If the editing studio permits it, we recommend bringing everyone who's involved in decision making with you when you cut your reel. This way, everyone knows right from the start why you made the choices you did, and all input will happen just once. People bring their agents, managers, friends, spouses, and parents (and sometimes even babies and pets, which is definitely not recommended). When deciding who to bring, remember that you'll need all the concentration you've got, and the longer it takes to make a decision, the more it will cost you. However, if you make your tape alone and then go home and show it, you'll discover that everyone you ask will have a different idea about the proper order of the scenes. The editor will not be there to help you explain the decision. You'll end up booking another session to rearrange the scenes. If your committee can't attend your edit session, get their input first.

When the advice of friends is important to you, be careful. No matter how close to you or how skilled they are, they are not paying for the tape—you are. It's not their careers that will be affected—it's yours. You are making a tape to get work—period. Any opinions should be measured against that objective. Listen to the advice, but make the decision yourself.

Some actors have a major problem dealing with the third party syndrome. They listen to absolutely everyone's feedback on their new tape, and after each critique, they're back in the studio. Everyone loves to be a critic. To give advice is free, and the ego lift for the advice giver is irresistible. Make your tape so it's suitable to yourself first, and those who have a vested interest (your agent and your manager) second, and let that be the end of it. The third parties can make their own tape and criticize that one until the cows come home. Don't keep rearranging the same material. There is a limit to what can be done. Put your money in a savings account.

7 The Montage

The montage on an actor's tape is a series of quick cuts used at the beginning of the reel to interest the viewers in watching the body of the tape. It is very similar in intent and structure to teasers at the beginning of episodic television. Like a commercial, it's a ten- to thirty-second spot of your best visual material, assembled in an intriguing way.

The montage gives you the freedom to create a wonderful image for yourself in a short amount of time. It's the most fun and creative aspect of putting your tape together. We're all used to seeing montages in commercials on television, and they get their point across in thirty seconds or less. This is exactly what you want to do.

DEMO TIP: A montage should not exceed thirty seconds in length.

Most often a montage is cut to music and the dialogue is eliminated. We have discovered that the maximum time viewers will watch before they realize they want to hear the person talk is 30 seconds. Even if lines of dialogue are interspersed, there is a limit to what can be learned from the montage. While this is a lot, don't hold up viewers who are ready to evaluate an actual performance. As with other conventions that constitute the visual language of television, we accept that most commercials will run for thirty seconds. Since our montage is unquestionably a commercial, we are likely to lose our audience at the thirty second limit because this is what they subconsciously have come to expect.

DEMO TIP: Your montage should involve the viewer emotionally.

We cannot stress this demo tip strongly enough. Engaging the viewers emotionally is what the entire industry is all about. The viewers, whoever they are, must be drawn to you emotionally for you to have an effect on them. In a movie theater, the doors are shut, the lights go down, and your image is on a towering screen. In a casting office, it is broad daylight, tension is usually high, time is short, and the job of your tape is to cut through all the interference.

Your tape must be constructed so that it grabs viewers emotionally or they will stop paying attention. The montage gives you a head start because you are able to employ music to increase the emotional impact of the visuals and set everything up for the presentation to follow. This is why we have devoted an entire chapter to the creation of the montage.

A montage is used for the following reasons:

- Along with your head shot, which should open the tape, a montage allows people to settle down and figure out who they're watching.
- It's an intro that will pique the audience's interest and encourage them to watch the entire tape.
- It balances your acting performance and shows your range quickly.
- It can help you break out of typecasting by demonstrating a range of characters in a short space of time using shots that could change your image (if desired), but may be too brief for the body of the tape.
- It can show close-ups and clips of scenes you like but can't use in the main section of tape for some reason.

We began doing montage openings to solve a variety of problems. Our clients often would have more material than they could comfortably use, and they'd say things like, "I'd love to use that close-up somewhere, but it doesn't fit with anything" or "I know my agent doesn't want commercials on my tape, but there are some great shots of me; I'd hate to lose them." We'd also hear variations on the following: "In all my scenes, I'm playing the bad guy. I do have one good guy role, but I don't say much. If I could get a few clips of that on my tape, I know I could break out of the bad guy rut." Other actors might complain, "My material seems so average. I need something really exciting to whip up the opening so they'll want to keep watching." You may be in this position: "I'm just starting out in the acting business, and I've only got one scene to use, but I have my modeling portfolio and a couple of commercials. Can I use them somehow?" A montage is ideal for all of these situations.

A well-structured, exciting montage takes time, and this makes it more expensive to create than the body of the tape. You may wish to use the fancy machines to cut the montage and a more basic system for the rest of the tape. This will save money in the long run. A thirty-second montage will take two to three hours to cut. It's time-consuming because the more cuts you make, the longer it will take, and not every image you choose will fit together smoothly and cleanly or have the emotional impact you desire.

Selecting The Music

DEMO TIP: Pick out music that has a good beat or a strong emotional pull.

The music at the beginning of your tape will help set the theme. Imagine some possible themes and what music would enhance them. Musical styles you might consider include: action, mystery, romance, high comedy, low comedy, urban, and country. There's classical, jazz, and even, as one of our clients successfully chose, music from the cartoons. "Thus Spake Zarathustra" (the theme from *2001: A Space Odyssey*) is somewhat overused for product demos (in fact we hear little else), and if you can find something fresh we recommend it. The right piece of music for your montage may already exist on one of your source tapes. It could be from an unsold pilot; a short-run, obscure, or unaired series; or an MOW in which you appeared. Television music is scored with the purpose of encouraging the audience to drop what they're doing and watch the show. Because this is precisely what you're aiming for with your tape, music from a television show is often a good choice, but it must be selected carefully.

DEMO TIP: Don't use music from a show you haven't appeared in, and don't use music that is too familiar.

"Hunter" has excellent music, but if you use it for your montage, viewers will believe you're a regular or a guest star on that series and will want to see those scenes. They might even search the tape thinking they've missed them. That does not leave a good impression. Another consideration is that if the music is from a long-running show, casting people are probably bored to tears with the theme. In that case, the music would be a bad choice for a montage unless your whole tape consisted of only that show.

Many people use a few bars of a popular song on their industry demo tapes. This is not legal, although we've never heard complaints when the tape was used only on an individual basis by artists and within casting offices. We recommend that you get permission before using this music or that you obtain your music from another source.

Most editing studios have a library of "needle drop" music, for which you pay a small fee every time they "drop the needle." (The terminology makes sense for those of us who remember phonograph records.) If you have friends who are musicians, perhaps you can talk them into writing something for you. Whatever method you choose to find your music, you will apply the same techniques in the montage.

DEMO TIP: If you must use music with lyrics, they should be sung by the opposite sex.

Preferably, the music you choose for your montage will not have lyrics unless you are the vocalist. The following story illustrates the problem. A client of ours decided to cut her montage to Cindi Lauper's "Girls Just Wanna Have Fun." We cautioned her about using a popular tune with a female voice singing the lyrics, but she had made up her mind, and we went ahead. The visual clips we chose featured the actor kissing different leading men, firing a machine gun, robbing banks, and making speedy getaways in flashy cars. Because the song was a comment on the visual action, the result was humorous. A few weeks later, after she'd circulated the tape, we asked her about the feedback. "The tape's doing great. People love it," she said. "I've already been hired for two jobs, but you were right about the music. Even though it's obviously Cindi Lauper, I still end up explaining it's not me singing in the montage."

DEMO TIP: The best music for a montage is the very beginning of the number.

The first twenty to thirty seconds of the song, the introduction, is often completely instrumental and comes before the singer begins the actual lyrics. The intro is meant to grab you and make you listen, much like television or film music. Therefore, it is often the best choice for a montage.

You may find that later in the song there's a few bars you prefer, like a musical chorus or a certain group of phrases. When you start to work with that part of the song, however, you'll discover it doesn't have the impact of the opening. It will also be harder for the editor to cut. Even if it sounds as if there's an appropriate break to cut in, most of the time there's some sort of cross fade, an inappropriate change of key, or something you have to struggle with technically. This requires extra time and forces you to spend more money, and after a period of trial and error, you'll probably give up and go to something else.

DEMO TIP: Supply your music on compact disk or cassette.

Used records can sound scratchy on a high-end system, and equalizing out the noise can dull the music. Tapes and compact disks (CDs) are generally easier to work with and will save you time.

The Editor's Multiple Tasks

You'll find detailed accounts of the editor's technical activities in Part III, which you may find to be interesting and helpful reading. For now, we'll briefly define various terms and functions to get you through the creative aspects of the montage.

DEMO TIP: Always lay the music track down first.

The editor will record your selection of music onto a preblacked, time-coded tape—that is, one that has previously been recorded with a video signal and with numbers on the frames to keep the machines accurate. It is especially important when creating a montage to have a time-coded tape because the time code makes each edit frame-accurate—that is, the edit will fall on the exact frame you choose—and the edit will never "slip." When using control track (little blips on the tape that the machine can read but which are not as accurate as time code), each time you preview an edit, the tape slips a little farther away from the point you selected. You will go over the same edit point many times for creative reasons, inducing wear at those points, and you don't want to get everything else right only to find the (non–time code) control track has slipped. This is hard on the tape, frustrating for everyone, and unnecessarily time-consuming. A frame of video lasts one-thirtieth of a second, and you will soon discover that one-thirtieth of a second makes a difference in a montage. You will often want to lengthen or shorten the music after the montage has been made, and by using time code, this can be done perfectly the first time. If the system you are using can only handle control track, redoing the music could take lots of time, and you'll probably never get it exactly right.

The editor may choose to record the music on a scratch tape, or work tape, first. If so, they will transfer the music from the work tape onto your new edited master. This is the time they are checking audio levels, possibly equalizing, fading in and out, and making sure that there are no pops or hiccups of stray sound. This is one of those times when little appears to be going on, but in fact the editor really needs to concentrate.

If you have decided to use some dialogue on the audio track, make sure that the editor knows this and records the montage music at an appropriate level. This will also tell the editor to record the music on a single track, leaving the other open for your dialogue. Keep in mind that if there is background music from the show combined with your dialogue, it won't mix with your montage music. If you want to go to the trouble of re-recording your dialogue in the editing studio, this can be done. We've used this technique to good effect a number of times. By limiting the dialogue to a single word or a very short line, it will blend just fine.

The Visual Concepts of Montage

To get some creative understanding of the technique of creating a montage before you go on the clock in your session, there are several things you can do. Start by searching out Sergei Eisenstein's book *Film Form and Film Sense*. Eisenstein was the Russian father of the montage, so have a look at his chapter on the subject. Then watch music videos and commercials that employ the montage, as most do, and study the movement in them. If you have some on videotape, watch them a few times with the sound off. Many rental movies contain montage sequences, and of course there are the teasers from all those television shows that are on every night.

DEMO TIP: When creating a montage, think in terms of choreographing a dance.

A good montage should be greater than the sum of its parts. A properly executed montage has its own emotional impact. In a dance, each step is chosen by the choreographer; it comes out of something and it moves into something. The dance has rhythm, flow, and style. The choreographer leads the eye around the stage, just as your montage will lead the eye around the screen. An individual step may not look like it's going to work until you see it in the context of the one that went before and the one that follows. It's the same with the selection of shots and cuts you plan to work with in your montage. In editing, you have the advantage of stopping and starting with ease and trying one or more steps at a time. If it looks jarring when you preview the cut, trim it (move it one or more frames), and try again. Preview the cuts until the eye flows easily from one point of focus to another.

You may plan a montage using only stills or only moving video or a combination. The important thing is that the montage continues to flow. If your eye gets used to moving images, a static head shot will break the flow. You will normally try to avoid breaking the flow, but in some cases you may have a reason to do just that.

The number of shots you will need depends on the style and length of your montage. Remember that you are not trying for a job as a montage producer. The thirty seconds of video you create must *feature* you, not upstage you with dazzling video footwork. The average for a thirty-second montage is a bit under two seconds per cut, and sometimes one second is right on the button musically. If you have twenty good cuts to choose from, that should be plenty. Of course, you can create a fine montage using far fewer, and there's nothing wrong with a ten-second montage. A shot that is flowing well can run over two seconds. Super quick cuts should be used sparingly, usually all in a group. Again, a combination of moving video, photographs, or freeze frames can be used. It's up to you.

Still Photographs and Portfolios

DEMO TIP: Not all of the close-ups or photographs you've chosen will work.

When incorporating photographs in your montage, you switch from one medium to another, which sometimes changes the overall impression of a shot. The aspect ratio of the screen is opposite to the standard 8″ × 10″ photograph and can entirely change the look of the photo. You need to frame each shot with the camera before you can tell how it's going to look. For this reason, you should bring all the stills in your portfolio to the session and line them up on the screen.

The camera's ability to crop a photo gives you a great deal of creative latitude. It can make close-ups more intense or frame out an annoying clump of unruly hair. It can also take another person out of the shot.

You may use special effects when shooting your stills. The special effects generator (SEG or simply *switcher*) combines various video sources, such as cameras, tape, and electronically generated backgrounds. It can put borders and frames around photos in any color. Your editor will be able to demonstrate these possibilities as you go. The next time you're watching a documentary on television, observe the way the stills have been handled.

With the digital video effects machine, you may be able to create a number of effects in combination with the switcher. Strobe, for example, is very effective: The camera appears to push in with a series of sharp jumps. Posterization or paint effects can give the picture an impressionistic look. A multitude of other goodies are also on line, and once again your operator will be only too delighted to show them off. Keep your eyes open when watching television to see how these effects are generally used.

Once again, discipline is required to limit yourself to certain choices, but it saves you money when it's time to write the check.

Before your editing session, prepare your photos (magazine covers and other professional photography) by removing them from your portfolio. Avoid using your originals. They could be damaged no matter how careful you are. The paper used for originals is stiffer than that used for duplicates, but it has a tendency to ripple and won't hang or lie flat, causing time-consuming reflections on the screen. If your originals have been retouched, the white cover fluid will reflect the light and result in a splotchy appearance. However, originals can be used if that's your only option. Color photos are fine to mix with black and white, and sometimes the SEG can be used to colorize black and white very effectively and give a variety of effects, such as a sepia tone or a soft, romantic glow.

When selecting still photographs to include in your montage, do not use the following:

- ordinary snapshots because they look grainy and unprofessional and are difficult to frame
- shots of you getting ready to do a performance
- amateur acting get-ups, such as warts, moles, and skullcaps
- pictures of yourself as a child (strangers find it somewhat embarrassing, and it doesn't express anything useful)

Professional stage shots are of questionable value. If that's all you have, use them judiciously. Unfortunately, stage shots don't convey anything concrete to people in the television and film industry. In fact, they could make agents afraid you'll overact if they cast you.

Still photographs need not be static. Camera movement adds interest to your montage when you mix stills with "live" video. You can pan up a full-length shot, zoom in, zoom out, and pan left and right.

You will save time by shooting all your photos at once on a work tape because it eliminates running back and forth to the copy stand or easel each time you need to lay down a shot. You will also benefit by being able to try a number of different camera moves for each photo. Record all the moves on the work tape, then see which work best as you mix them into the montage along with the live video. You can try different zooms, both in and out, and at different speeds. Pans up or down on a full body shot work well, and you can try creative ideas, such as starting on the flower you're holding and gently widening out to reveal your wistful smile. Once you start working with the photos, all sorts of ideas will come to you.

If you try to shoot each photo "live" on a system that cannot read time code or with an editor who is not experienced enough to use it, you also run the risk of ruining your timing. For example, if you want a photo to come in on a particular beat of music, the editor will need to back up into the previous picture by one or more frames to make sure there's no gap or dropped frame (a frame of leftover video) once the edit is made. Because you're shooting within an area as small as 8" × 10", the camera movement must be precise. If the editor must redo a right-to-left pan on a photograph more than a few times, you start to carve frames off the previous cut. Remember, your clips are less than two seconds, and with only thirty frames to a second, every frame counts. The next thing you know, the edit point has started to back away from the beat of music where you planned the picture to come in. Pretty soon you will get the picture first and then the beat. If you shaved too much off the previous shot, you'll have to go back even farther and edit the previous shot back in, doubling the wasted time.

Another problem with shooting directly onto the edited master is that the editor must coordinate the start of the camera move with the exact moment of the edit. This is just one more thing that can go wrong, requiring you to go back and try again, redoubling the time you're spending. When you use a work tape, you will go down one generation, but the minuscule loss of resolution will never be noticed. If you do not use a work tape, you once again run the risk of wearing the oxide away from the in point (the place where the new edit starts) and causing a physical weakness that may be unnoticed on the master but can show up as a glitch on your copies. The only way to fix it will be to do the whole job over again on a new, unspoiled master tape! This potential occurrence is one reason you should buy only broadcast tapestock for your master. Broadcast tape is specifically designed to withstand multiple edits. If you are mastering on S-VHS, use professional-grade tapestock.

DEMO TIP: Begin the montage with your current head shot.

Begin your montage with the head shot that motivated the agents or casting agents to see your tape in the first place. Give it about four to six seconds, letting the beat of the music dictate the duration. Watch carefully to make sure that four seconds isn't too short, or six seconds, too long. You might want to start the head shot in silence and then bring up the music, or vice versa. If you're cutting to the beat, let the head shot and name remain for the four seconds, then choose an interesting point in the music and cut to the beat from there. Finally, when cutting photos shot with camera movement, vary the direction and type of move to avoid boredom and repetitiveness.

The Moving Shots

When you choose your moving shots, remember the image of the choreographer. Whether they're close-ups, two-shots, action shots, or whatever, study the movement on the screen. These movements are the steps of the dance.

DEMO TIP: Keep the movement moving.

When creating your montage, you are the choreographer and it's your job to keep the dance in motion. Let's say you have a close-up that starts with a pause because you're listening to someone, then you throw your head back to laugh, and you bring it down again and pause once more before the scene cuts to something else. Start your shot at the precise moment that you begin to lift your head up, and cut it at the height of the laugh, which should also be the highest point of the movement. Now look for a shot that takes the viewer's eye in the opposite direction. A shot where you're on a toboggan sliding down a hill with a couple of kids would make an excellent juxtaposition. If there's a pause at the beginning of the laugh, it will be boring. If the toboggan is too far away for us to see what you're doing, it will be boring. "Just right" will elicit cheers in the edit bay.

DEMO TIP: A thirtieth of a second, one single frame, can make the difference between magic and boring.

Since you've just used unrestrained movement, why not try a sudden stop? You may have a shot where you're heading for an elevator and the doors shut almost on your nose. You turn to the camera with a look of exasperation. Now play with looks of exasperation for a while. You're sitting in a car as a cop writes you a ticket; you can't get the last drop out of a bottle of ketchup (from a commercial you thought you'd never use). Next, a wide shot of you hauling on the reins of a stubborn donkey. Since we're in an agricultural mood, how about three quick stills of you in various western get-ups, then two seconds of that scene with you sitting on a rail at the rodeo smooching with a famous actor. Now your whole series of images is linked together by movement as well as ideas.

A Simple, Inexpensive Montage with Big Impact

Use moving video only. Scan through your material at home, and pick out ten close-ups. Don't choose shots that might mean trial and error when you're putting them together in the studio. If possible, select shots with just your face full on the screen. Make sure each is somehow different in emotional tone. Aside from your 8" × 10" head shot, don't use any still photographs. For music, use the title music from the first show that will appear in the body of your tape. Make your cuts on the beat rather than action to save trial and error time, then cut straight to the title of the first scene as the music continues, end the music, begin the scene. This is the most cost-effective way to create a montage that has impact.

DEMO TIP: Design the montage around the best material you have.

You must be happy with every frame on your montage, and nothing must make you feel less than 100 percent enthusiastic about it. As with the tape itself, if there's something about a look that makes you wonder whether you should use it or not, the answer is to drop it. Many demos suffer from padding—that is, employing images that don't work or that don't do the artist justice. Think about every shot. Are your purposes served or defeated? If it isn't working, cut it.

Finding a Theme

DEMO TIP: Find a theme to work from.

The theme for your montage can be almost anything: a musical theme, a certain type of movement, or a feeling. You may discover that you wave your arms a lot or kick things a lot. You can use this as a theme. In one shot you kick a door shut, in another you kick a door open, in another you kick a football, and in another you kick a tin can down an alley. Intersperse these throughout, so it goes like this: a close-up where you quickly look over your shoulder; you kick a door closed; you wave your arms as a bus speeds by; you kick a football; you kick a door open because you're carrying groceries; you're in court and you throw a pencil in the air; finally, you kick the tin can down the alley ahead of you as the shot fades out.

Another possible theme might be a mood, like sultry. Select some sultry music, and use languid camera movements for your photos. Choose moving video close-ups that are soft and slow. You might smile next to a glowing candle, slide gracefully out of a car, do two seconds of a waltz, then blow a kiss. When making a transition from photos to moving video, it's possible to do dissolves even on a cuts-only system, if a rudimentary switcher has been installed. Dissolves should be used to emphasize the soft, sultry theme.

DEMO TIP: Cut so that the action occurs on the beat rather than place the cut itself on the beat.

An important secret of the art of montage is to edit so that the action occurs on the beat. Cuts on the beat are solid, less subtle, and can produce a military kind of feel. They are definitely useful in certain instances. When you're changing the subject, for example, you can emphasize the change by making the cut on the beat. As an illustration, let's assume we're doing a montage for a commercial cinematographer with plenty of action shots. Baseball connects with bat on the beat; car door slams shut on the beat; cap comes off bottle on the beat; bathing beauty turns head on the beat; plane's wheels touch down on the beat; girl on swing reaches apex on the beat; and twins pop gum in their mouths on the beat. That's all there is to it.

These cuts must be as precise as you can get them; if one's a little bit off, it will show. Should you begin to get confused—and everyone does—go back to the beginning of the montage and play through to the spot where you're having trouble. If you can't find the exact moment, play through the montage again.

You can get creative by studying the music. If it gets tinkly, use the china shop shot; if it becomes thunderous, use the cut where you tackle the ball carrier; or, for fun, do it the other way around. Try the shot with the light kiss at the point where the music goes "ping."

DEMO TIP: Carefully look for humor.

Find humor in the shots and in the juxtaposition of shots, but be careful to avoid anything vulgar. If you see some humor, you must be certain that it will be amusing to everyone, just like they try to do on nighttime television. Jokes and gags are as dangerous in the montage as they are anywhere else in your tape, so use your discretion.

Don't forget to color correct your montage. When you're using quick cuts from a variety of sources and you're busy thinking about other things, like movement, theme, and the rest, you can forget to watch the changes in color that occur when working with tapes from different sources. As the shots are cut together, make sure the hue is adjusted for each one and that the faces are all flesh colored. (The cheapest systems do not have this capability.)

Identifying Your Goal

DEMO TIP: Decide what you want your montage to achieve.

Identify your goal, and use your montage to achieve it. Let's suppose you're typecast a lot. You've finally said enough's enough with the guns, and you're determined to move on to leading man roles. To achieve this, you need casting people to see you in a different light, but you have no good guy scenes.

Look through your commercials, your head shots, and other work that you don't plan to use in the body of the tape. Try to find some vulnerable close-ups. Because it's a montage, the viewer won't see that your character has just killed someone when you cut to your hurt and puzzled close-up. You may be just about to say something sleazy, but for a brief second you appear perceptive and intelligent. When you're logging your material for the montage, try to see these shots out of context, and you'll be surprised at how much you have that can work to change your image.

You may want to alter the character balance of the reel for other reasons. This is especially true for people who play psychopaths, murderers, and sociopaths. With these types of roles, it's possible to be so good at your performance that it works against you. Why let casting wonder, even for a second, if you could be trouble on the set? The montage allows you a chance to comment on the characters you've played, letting the real you show through.

For women, the airhead role is a big typecasting problem, and the above techniques can be used to counter it. Find shots that make you look like a doctor or a mom or a helicopter pilot. Even if they're only a second long, once they're cut together, they will reflect your new image.

Using Variety

DEMO TIP: Go for variety.

You hope the person watching the montage will say, "Wow, they've been everywhere!" Different costumes, different hair styles, different locations, different attitudes, day shots, night shots, and different costars are all tools for you to use. Try not to repeat anything, unless it's done intentionally. Keep viewers guessing about what comes next, and they'll never notice the time.

Ending the Montage

The last shot of you that viewers see before the body of the work is a good place for a pleasant photo or freeze frame. Look for something that leaves viewers with a positive feeling. Here is an opportunity to use that 8″ × 10″ head shot you really like but are not currently using for submissions.

Montage Alternatives

The One-Liner Montage

If you have a lot of one-liners that get applause, big laughs or are just funny by themselves, try stringing them together in a montage. This can be a good way to utilize "under fives." If you've often been cast in the under-five-line category, now is your chance to put those "under fives" to work. Edit together a one-line montage of about thirty seconds, then use your very best (and perhaps only) scene.

You might be able to use the one-liner montage at the end of your reel, for a specific reason. One of our clients appeared as a semiregular guest on a sitcom, and had a few (very funny) lines on each show. When she was finally featured in an entire episode, the plot was a serious one. Her problem was that the large scenes were serious and the funny one-liners couldn't be cut to play like a full scene. We decided to use two well-cut, serious scenes up front, then a montage of her best one-liners with plenty of laughs and applause. Not only did casting people see the dramatic side of the actress, they also saw one-liners that were delivered with real punch and received great reaction from the audience.

An Alternative Opening

If you've been on a talk show, let the host introduce you. It's an impressive way to open your montage, and it immediately lets a viewer see the level that you've achieved in your career. Use the intro only. This should precede your head shot. In this case you might save the head shot to end the montage. Once your name is announced, you're seated, and the applause is dying down, begin the montage or the body of the tape.

If Gene Siskel and Roger Ebert raved about your performance in a particular film, by all means use that as an opener. (Do not, however, use the review you got in the college paper when you were the youngest actor ever to play King Lear.)

Using Shots of Other People

If you've worked with many different stars in the past five years, you can make a montage that shows you sharing scenes with them. This tells the casting agents that you've been around and can hold your own in the big time. It also helps your agent negotiate more money!

DEMO TIP: Make sure others' attitudes support you.

If you decide to do this, which may only take up a short segment of the montage, you can benefit from their character's attitude toward the character you're playing. They might be saying or appear to be thinking some of the following: "I love you," "Isn't she great?" or "Isn't he heroic!" They might hold out their hand to you, look up to you, or even just smile at you. This can make your montage quite funny, depending on how you structure it.

Ending Your Tape

Many people ask about doing another montage for the ending of the tape. From the reaction that we've had to these, our feeling is that it's best to keep the end of the tape very simple. If viewers have watched this far, it can be overdoing it to insert another montage, unless it's for the kind of specific purpose mentioned earlier. A strong ending shot and a freeze with your name superimposed is all that's necessary. You walk over to a door, and just before you exit, you turn and smile. Freeze frame. Super name. Music up. Fade out.

8 Updating Your Tape

This chapter may be brief, but it is one of the most significant in terms of saving you money over the long haul. You can save as much as a couple of hundred dollars per reedit session if you apply the suggestions offered here.

When to Update Your Tape

Once you have gone to the expense of creating your original master demo tape, it should last you at least two years, often significantly longer. Update your tape whenever you have done something that could advance your career. Don't bother to update it every time you get an under five, particularly if you already have a similar scene or character on your reel. Before you decide if you can afford to update, remember that you will also have to call in all your copies and rerun and relabel them. (If you do, be careful to include the new date on the label so you can tell the new copies from your old ones.)

When you land an important new role you wish to add, ask the editor to copy the edited master of your demo reel to a three-quarter-inch (or S-VHS) submaster and to insert your changes on that. You can change shots in the montage and rearrange scenes as you go.

Generation Loss

Until recently, making copies, particularly in an industrial or off-line edit bay, lost so much of the picture quality that going to a submaster and then copying again (thereby making a copy of a copy of a copy) produced unacceptable results. This is why generation loss became such a concern. People actually got emotional about it, and some still do. However, the machines and tapes in use today are so improved that generation loss is far less significant. The submaster loses a little picture quality, but this is usually unnoticeable. The use of a submaster saves so much money that it makes regular updating feasible. As long as your source material is good enough to begin with, you'll be fine.

It's a good idea to check for generation loss before going ahead with an entire session. Simply make a copy of a portion of your original edited master to the tape you will use as the submaster, then copy that portion to a VHS, which represents going down the same number of generations as the final version. Play back the VHS, and if it looks good to you, go ahead and make the submaster. If the VHS looks grainy or washed out, if the colors bleed, or if the picture flips, there is not enough signal to go down the extra generation. You'll have to reedit those particular scenes from scratch. This does not happen often, however, and it may only be a problem on one or two of your oldest source tapes. In that case, edit what holds up onto a submaster and redo the problem sections from scratch.

Updating the Montage

When updating the montage, think back to the mood and theme you had in mind when you created the original. Very often people insert new shots into the montage that glaringly break its mood and flow. Remember, the montage is an entity greater than the sum of its parts, so treat your original work with the respect it deserves.

Editing with a Submaster

Copy your master tape, keeping the montage and the other scenes that are in place where you want them, then add in your new material from the source tapes in the order you've chosen. It happens like this: The opening, the montage, and the first scene remain as they are. Then comes the new scene you want to add. After this, go back to the master for the second scene (which now becomes the third scene on the submaster). You continue recording from the master until you're halfway through the scene that follows; there you stop because you want to eliminate some dialogue that you now realize doesn't really feature you. Scan past that dialogue without recording, and pick up the scene with your close-up, being careful that the audio levels match. Now continue recording the rest of the scene from the master. Next, pop the master and switch to the air check of your other new show. Edit in the scene you've chosen from it. You will do the final editing of this new scene as you go along. You have decided the rest of your tape should remain the same, so pull the air check from the source machine, reinsert your master, and record to its end.

9 Demo Tapes for Comedians

Tapes for comedians are quite different from tapes for other performers in a number of respects. A montage is usually unnecessary unless your act includes different costumes and characters or you perform some type of circus routine along with the comedy, such as juggling, blowing huge bubbles, or doing magic.

Comedians often need two demo tapes: a short tape, which will go to television casting agents or serve as a calling card tape, and a long one, which will go to the club managers who want to review your entire act before they hire you. For the latter, string together one, two, or three of your best routines, and keep the editing to a minimum. These tapes, although long, will not be too costly to produce because little editing is involved. The material will usually come from videotape shot in one of the clubs you've played. Many clubs have video cameras already set up, and they allow you to purchase a copy of your act after the show. If there are no video facilities, you will need to hire someone or to bring in a friend to tape your act. Remember to follow the rules on live taping in Chapter 12.

For the shorter demo tape, your source material will come from air checking your act if you're doing television or from shooting your standup in comedy clubs. Most comedians' tapes include a bit of both, and the clips should be carefully placed to be effective. Knowing what material to put on the tape is a little easier for comedians than for actors: It is gauged by audience reaction.

DEMO TIP: If they're laughing, consider it; if they're not laughing, don't.

When mixing both television and club performances, keep some television work up front, put the club performance second to last, and save a very upbeat television clip to go out on. This is a little different than our suggestions to actors. The reason is that the lesser technical quality of the tape shot in a club may not have the impact of real television, and you want to leave viewers with the image of real television if you can.

As with actors, if you've been on the late-night talk shows, open with the host's introduction before your entrance, keep the handshake or greeting, then cut to your act. If you have a number of performances to choose from, select clips where you're knocking them dead.

A demo tape only needs to be three to six minutes in length to do the job, although a slightly longer tape is okay if the material is really hot. Still, it's to your advantage to pare it down to your very best work. Keep it tight, fast, and leave the viewers breathless.

When editing, you can keep costs down by selecting a cuts-only system instead of an A/B roll system. A comedy tape should be as upbeat and snappy as possible, and you can usually get away without dissolves. If you do choose an A/B roll system, you can make any jump cuts more palatable by making the transition with a dissolve or other effect.

Controlling the Audio

When shooting a standup routine, you must pay special attention to miking, and, if possible, both you and the audience should be miked. The video operator can sometimes "take a feed" off the audio board by running a cable from "line out" to "audio in" on the recorder. Line out carries a signal strong enough for the VCR to record. To take a feed simply means to plug a cable into another piece of equipment and then connect it to your own machine, rather than duplicating microphones and other equipment. If you are planning to mike your routine yourself, be sure to consult the audio section in Chapter 12.

DEMO TIP: The more equipment a signal must pass through, the more opportunity there is for something to go wrong.

The most effective way to pick up the audio in a club is to run the comedian's mike directly into mike channel one of the recorder and the audience shotgun mike into mike channel two. With this setup, you will have much more control over the laughter when you go into the studio to edit your tape. The editor can ride the gain on the laughs, enhancing the audience response considerably. Camcorders with only one mike input will not have this capability, but the audio could be mixed by using a Y adapter, which joins two cables into one. Be careful to test this setup before you begin to make sure you are not being overpowered by the audience.

If you need to make a cut using either audio configuration, there is a simple way to make the transition smooth. Make your cut, then start the incoming audio at a slightly lower level, fading it back up to normal. This will mask the change in audio enough to make it palatable. You will still have a jump cut in the video of your performance, though. To cover this, simply use a reaction shot of the audience. Clubs with a video setup always get these reaction shots, but if you hire someone, have them pick up the audience at a couple of appropriate moments so you'll have something to cut to when you edit. Decide beforehand where the camera operator should do this so you don't ruin your best joke with a camera move. (If you're shooting with an invited audience that you can control, at the end of your routine turn the camera around and re-do a few jokes. The audience really gets into it as a rule and you'll have some great shots.)

If this isn't possible, you may wish to mix a bit of audience laughter behind the reaction shot, which will mask the jump cut even further, but be careful that the laugh mirrors the audience's expression. If they're all sitting there stony faced and you plant uproarious laughter on the track, it will not generate the viewer response you were hoping for.

Even if it's not possible to run a second audience track when you shoot the performance, you can still enhance the audio to some degree. This is time-consuming and very exacting work, but the results can be worth the extra time and cost. Ask the editor to go through the tape and collect on a scratch tape audience laughter and reactions where you are not talking. Then edit your master tape, using only one audio track for your performance. On a second pass, find appropriate laughs and reactions on the scratch tape, and carefully add them to the second track so that they blend. This can work extremely well.

Two of our comedian friends each did the same show on different nights, and each got a tape of his act. The first comic had great luck: The audience was practically all college students and tourists. The second guy didn't have it so good. Even though his performance was super, the management had bussed in a whole audience of senior citizens who didn't let out a peep through the whole act. Since both tapes were shot in the same studio, we were able to make a scratch tape with laughs from the first comic's performance, and, with his permission of course, graft his laughs onto the second comic's tape. No one could ever tell.

Commercially produced laugh tracks are not useful for this purpose. They really sound like what they are—canned laughs—unless you're mixing in a very high end sound lab.

If you are considering inviting an audience for your taping, by the way, you may want to get a few fellow comics together, book a room, and do a whole evening. It will be more attractive to the audience and your costs will be considerably defrayed. It would even make bringing in a higher end production crew more feasible.

10 Demo Tapes for Other Entertainment Professionals

These days we hear the words "I need to do a tape" from an ever-widening circle of entertainment professionals. Competition is intensifying as budgets shrink but, as always, those who do the hiring must regularly choose their production team from among new faces. Unless they know the individual personally, the video demo reel is the most convenient and reliable way of making a decision concerning an applicant's qualifications.

For the job seeker, it is a great help in demonstrating a wide range of assignments in a manageable space of time. Those who have discovered the advantages of a video demo or sample reel will never allow themselves to be without one, and they are also careful to keep them current.

Cinematograhers

Cinematographers can put together their tapes in much the same way as performers.

Most cinematographers have access to one-inch or Betacam suites, and you will probably create your master on one of these formats. Contrast is a problem when going from film to video, and it's important to retain as much of your lighting quality as possible. The major difficulties will occur in the dark areas, and there isn't much you can do but curse video. Watch color correction like a hawk when editing from different source tapes. If you're editing on three-quarter-inch, the Sony SP machines retain the most quality in terms of generation loss.

The length of production demo tapes is not as restricted as the length of actors' tapes. The main rule is to keep things moving so the viewers don't think they've seen everything you do before the tape is over. We know of excellent six-minute reels, but eighteen minutes seems to be fairly standard. When you're approaching half an hour, it's getting long.

If you're a beginning camera operator, shooting your own demo reel to apply for film school or a job with your local television station, the key word is *variety*. (If the school you have in mind has written guidelines, follow them to the letter.) Shoot some news-style footage, a product demo, an actors' scene, and so forth. Watch television with the sound off. View the television screen as if it were a viewfinder, see what the camera operator or director of photography (DP) was trying to do. Grab your camera, go out, and try it yourself. If someone offers you a job at a television station banging flats together or pulling cable, take it on the "foot in the door" theory. You can probably get access to a broadcast camera, free advice, and a late night editing suite.

If you just purchased your first camcorder, remember everyone has a first day with a new camera. If you're on rung one and hope to be DP on a major feature film someday, there are two words to think about: *quantity* and *quality*. Shoot a lot, but shoot carefully. If you just run around shooting everything in sight, your concentration will begin to wander and you'll stop learning. If you feel exhilarated by what you shot; if you planned it, either before or during your shoot, and it worked; if you feel like you've just got to show it to someone, this is a sign that you are succeeding.

Choreographers and Dancers

Since the advent of music videos, choreographers have been very much in demand. Even when dancing is not involved, they are often on the set as "movement stylists."

A montage of thirty seconds or so always works well for both choreographers and dancers. It's tricky to insert moves from different dances over the music you've chosen, but it can also be impressive if it works. It's a quick way to show the artists with whom you've worked and to give producers an idea of your range and style.

Six minutes is a good length for a tape, but there are no rules. Your tape can last twelve or fifteen minutes or longer if you decide to include whole dance numbers.

Television, commercials, and film can be effectively used for a choreographer's tape, and stage performances can be used with discretion. The production values of videotaped stage performances ought to be high enough to look fully professional. Judge a stage piece carefully before deciding to include it on your reel. The same discretion should be applied to the selection of industrial work and sales promotions. Remember, if you're uncertain about whether to use a particular clip, drop it. You want to be so proud of your tape that you can't wait to show it again and again. If you have a very good piece that was shot on consumer video, place it second to last, and cut it down to include only the very best part of the clip. Conclude the tape with something really slick that leaves a lasting impression.

In planning your tape, remember that its job is to get you a job. If using only one dance number and running it all the way through will give you the strongest demo, there is no harm in doing just that.

If you want to cut together highlights from a single number, lay down the music, then recut the visuals to it. It can be an effective showpiece.

If an A/B roll system is available, you can dissolve the picture and cross fade the audio to make your transitions. If effects are available, try a mosaic pattern for transitions; fade out the old audio, then fade in the new. Fading to black and back up again should be used sparingly, as it tends to suggest that the show is over. Try making a jump cut while you cross fade the music, or use some of the above effects as transitions instead of fades to black. Fades to white are becoming increasingly popular and avoid suggesting the show's over.

Titling each piece can tend to slow things down, so if you choose to do this, keep the pace upbeat and snappy. It's possible to superimpose or generate a card between each number, perhaps beginning the music under the card. One interesting idea is to super your title in a lower corner à la music videos.

If you have considerable stage credits but no video of the work, try making a credit role of all your shows, and scroll it at the end of your tape over a snappy freeze and some great music.

An idea that is extremely effective is to obtain electronic press kit (EPK) footage from the on-set videographer. The footage can be used in many creative ways, depending on the type of coverage you have. You can put together a little trailer showing you actually choreographing the dancers or as a dancer rehearsing the moves, then cut to the final product. If the video camera operator would like to do a short interview with you, make time. Afterward, brief clips of this can go on your demo tape either as a voice-over or an on-camera appearance.

Talk show interviews are another source of interesting material. As with actors, open the reel with an introduction by the host, but add a few words from the interview that comment on the visuals in appropriate places. Comments on the tape are more useful for choreographers than for dancers. Dancers' tapes should focus on the actual dancing.

When getting ready to make a tape, be sure you are about to produce the type of tape that satisfies your agent. Find out what your agent likes and doesn't like before you spend your money in the editing studio.

Child Actors

Tapes are as important for child actors as they are for the rest of us. When creating a demo reel for a child actor, bear the following tips in mind:

- Keep the tape current. Children change very quickly, and if you send around an old tape, you'll be heading out to the wrong auditions.
- Don't forget to air check all the child's performances. The one you miss is the one you'll really wish you had.
- Since you'll be updating the tape regularly, there's no harm in keeping the tape quite simple to keep your costs down.

If your child star hasn't worked in a while and has outgrown his or her category, replace the head shot at the beginning of the reel with a current one, and consider adding three or four composite pictures—as long as he or she doesn't look too posed or characterish—so everyone will know what the child looks like these days. You might consider investing in a video screen test or having the child do a brief video interview for the sole purpose of letting the agents see how the child has grown.

Game Show Hosts

Of all the different categories of performers, the tape for game show hosts should be the snazziest. Open with a brief cut of the opening graphics, your card with the theme music running underneath, and the voice-over announcer naming the show and identifying you as the host.

Six minutes is long enough for a tape that features one show; nine minutes is plenty for two shows. If you have a number of shows, a couple of minutes on average for each one will work very well.

Choose your clips in response to the following questions:

- Am I in control?
- Are my witty remarks and comebacks pleasing, charming, or funny?
- Does the clip show contestants at a high emotional level?
- Does it generate an immediate emotional response?
- Is there a good joke on me that I handle with aplomb, such as a Freudian slip or a silly flub?
- Is my energy level high yet controlled?
- Are the contestants open, happy people with pleasing and diverse personalities?
- Am I showing people winning?
- Is there a funny segment of a humorous loser?

- Am I at ease handling product names?
- Is there an obvious rapport with others on the show?
- Is there an obvious rapport with the audience?
- If children are on the show, are they responding well?
- Are there clips of my interaction with the audience?
- Have I come across to the folks at home?

A cuts-only system is perfectly adequate for this type of tape. The main problems will be in matching and cross fading audio. The audio editing suggestions for comedians in Chapter 9 will be helpful to you as well. It is important to review your shows at home and to log your tapes before the edit session. Narrow things down to the very best material you have, and work out the order. This will cut costs dramatically. If your career spans a number of years, open with recent material, add older clips in the body of the tape, and close with recent material.

For those who hope to become game show hosts, a video screen test is worth considering. Book a studio that offers multicamera shoots, and find two or three friends who have been or would like to be game show contestants. Next find a script of a game show that is no longer on the air and is very simple in concept and set design. Obtain three upscale-looking podiums, block the show with as little movement as possible, and shoot your screen test. Edit six of the best minutes onto a tape, and go land yourself one of the most high energy, intense, and unpredictable jobs in show business. Don't forget to let your two or three would-be contestants have copies of the tape. They can submit them when interviewing to be contestants on real game shows.

News Professionals

News Anchors, Sportscasters, and Weather Reporters

Assemble a brief tape of your very best moments. Start with the largest stations you've worked for, then add others. Look for moments that reveal your personality in ways that attract the viewer.

Consider including the following:

- your coverage of an important or a tough event
- a shared moment with a colleague or the camera
- your handling of an unexpected situation on camera
- a vulnerable moment
- your easy, friendly, and sincere air of authority

Weather reporters have somewhat more leeway, as Steve Martin pointed out in *L.A. Story*.

Newscasters and Reporters

For newscasters and reporters it's important to save all your work exactly as an actor does. Fortunately, you never have to air check anything. The network or station you're working for will transfer your story from one-inch directly to a compilation reel or, more commonly, to a used short-run tape. Working newspeople usually have about fourteen cassettes to edit from when making their reels.

The station will transfer the story for you right after it airs. All the titling will be in place, including your name, the location, and the station logo. Make sure this is done while the news story is fresh and at hand. Because your clips are dubbed in-house, the generation loss will be minimal. Needless to say, your knowledge of videotape and editing is already extensive because you work with tape and editors daily.

Preselect a series of news bites before editing. You can obtain feedback from the news director or station manager about which items represent your best work, and these should form the core of your demo reel.

The stories are of secondary consideration now but the closer you can come to the structure of an actual news broadcast, while featuring you, the more impressive your presentation will be. Stories should be balanced between emotional impact and those that impart information. As with other tapes, the tape should begin with your very strongest piece of work. You will rarely use an entire piece unless it's exceptional or very well known.

If you have access to an A/B Roll system, choosing border colors and wipes similar to those used by the station will help to unify your tape and give it a fully broadcast look. Use the CG to come as close as you can to the station fonts when superimposing names, locations, and dates if you have cut around the originals in your editing.

News Readers

With the swift development of cable in the eighties, many actors began to be hired as "news readers" and "stock market report readers." The major requirement for these positions is the ability to convey information in a pleasant and confident manner. This is where putting together a demo gets tricky. What you want to do is to keep potential employers interested in your performance. The difficulty arises because the information is sometimes quite dull or is of interest only to a small audience.

Use the opening of the show with all the identifiers—name, show title, etc.—then lead the viewer through your best and most interesting pieces. Be sure to cut when you've seen enough of a particular segment. Make internal cuts to pick up the pace and maintain intensity. Close with something snappy and memorable.

A cuts-only system is fine for this type of tape. Spend a little time with the first opening graphic or the teaser and the voiceover announcement of your name. For the clips that follow, a station graphic is optional but not necessary. If it's in the shot, great. The most important thing is to get as much of you on the screen as possible in the shortest period of time. The tape should come across as upbeat, direct, and packed with news.

If you decide to shoot a video audition for a news job, keep your set simple, and if you have a choice, let it be news-like. The most important factor is that you know your copy forward and backward. Even if there's a Teleprompter, memorize your copy. Forget about using cue cards because your eyes have to move too much to see them.

Broadcasting Students

If you're fresh out of school, you'll have on-camera classroom work. Very often, everyone who graduates with you will have the exact same thing: the same stories, the same set, and most likely the same order. It's a good idea to take the initiative, as one of our clients did, and present future employers with something different. He was an intern at CBS for a semester and joined forces with another intern studying to be a camera operator. They shared the cost of a rented camera and portable recorder, bought some tapestock, and went on location. They chose three dramatic backgrounds, and our client reported on three very different stories in their entirety.

The result was a topnotch professional demo tape. The big plus was that it was different from all the other demo tapes prospective employers would be receiving from the other broadcasting graduates. In our editing session with him, we supered his name, with the requisite underscore, over the opening of his pieces. Because it mirrored the field reporters and actual news items we saw every night, it had the look of the real thing. He was pleased, we were pleased, and he submitted the tapes and got a job.

If the courses you took didn't supply you with any on-camera classroom work to use on a demo, you've got the freedom to make up your own. Call around to small video studios to find one with a news-type set and a Teleprompter. A three-camera setup would be best, and the cameras should be high-end industrial or broadcast. (Review the section on choosing a studio in Chapter 12.)

To keep costs down, get together with other students and see if you can book enough hours to qualify for a discount. Match yourselves up in news teams, which should always be male/female. This is the simplest and the most cost-effective way to start a news demo for yourself. You have more control over hair and make-up in a studio than on location, but having both types of work on your reel is highly recommended.

One client of ours went overboard, but it worked for him. He borrowed a huge sum from the First National Bank of Mom and Dad, hired a broadcast crew, and lined up volunteers to be interviewed in a variety of areas. A doctor talked to him about a controversial drug, a supermarket manager discussed computerized check-out counters, and so on. He brought us the tapes and we edited the whole thing into a completely real-looking news magazine with all the bells and whistles you'd find on television. When it was all put together, he went to a printer, had a press kit on himself made up, ran a zillion copies of his tape, and sent one to every station in the country. Yes, he did get a job, but we suspect he's still paying off the loan.

Magicians and Other Activity-Related Performers

Put your most professional material up front. Use an introduction from a familiar show and go quickly to the act so booking agents see what you do. A montage can be very effective, such as one card trick after another. Do cut-aways to an amazed host or an amazed audience. If Johnny Carson is watching your card tricks, let's see that moment where, bewildered and impressed, he does a take to the camera and exclaims, "He's good!"

A trapeze artist might choose a series of cuts letting go of the bar. You may want to cut in a bit of crowd reaction or show a clown looking up at you, watching intently. All reasonably well-shot video is usable, and footage of rehearsals can be as interesting as performances.

Whatever your routine, it's important to let as much of your real personality show through as possible. Very brief clips of interviews, either as voice-overs or on camera, are useful; they can be as short as one sentence or even just a few words. Behind-the-scenes observations are always interesting and help connect you with the viewer. Try to engage the viewer emotionally whenever possible. One theme for a tape of this type might be "How could anyone that skilled be that likable!" That one should be easy, right?

Stunt Performers

If you're a stunt performer, you are very lucky because action shots are perfectly suited to montage. Opportunities also exist for a humorous approach, such as getting beaten up in a number of shots, then turning the tables, or vice versa. You can freeze frame at the apex of a stunt, use strobe to good effect in a building fall, repeat the stunt several times (if it's interesting), or keep coming back to the same stunt (if it's interesting).

In a stunt involving several performers, it's possible with some switchers to highlight you in the mob scene, or you could point yourself out with an arrow, as long as the arrow appears to be in fun.

It's also permissible to describe the stunts on the audio track. For the most natural delivery, talk onto a scratch or work tape without using a script, and let the editor cut the good bits—called *sound bites*—together afterward on your master. If you're uncomfortable with a microphone, forget about doing a voice-over and just cut to music or use the actual audio track, whichever works best for you. Try using the title music from one of your films and editing the stunts to that, particularly if it's action-style music.

It might be possible to cadge some footage from the electronic press kit crew. They are usually around on stunt days. Be sure you obtain the right permissions. Behind-the-scenes footage of stunts is always interesting, even to people who spend their lives on sets, and the footage can be a trigger for conversation in an interview. One of the reasons we included a shot of a car wheel spinning in snow on the production designer's reel we described was so she could say, "Do you know what I used for snow in that shot?" and after a rhetorical pause, "Instant mashed potatoes."

Film Industry Craftspeople

Most craftspeople—such as greens people (they handle all the real or phony trees), set builders, caterers, animal trainers, and cockroach wranglers (yes, that is an honored profession in the industry)—can benefit from using a tape in their continual search for work or contracts.

To bring a consumer camera to a location is a delicate matter, and it's important to check with the assistant director (AD) and the production office a few days in advance. If they understand what you're doing, you'll probably get permission, and if they don't allow it, you may be able to take a series of shots with your still camera and edit them onto a tape with some movement (pans and zooms) to make them interesting.

You can certainly videotape your own shop or hire someone to do it. Tape whatever makes your job unique, and let the camera see how good you are at doing it. Be personable and professional. Let the prospective employer see what a valuable contributor you'd be on their shoot. If you plan to do the video yourself, refer to the information on camera work in Chapter 12.

Screenwriters

Generally, writers do not use video as much as they might to promote themselves. Writers who do use a demo reel have found it a very effective form of introduction to their work. They select a few of their best scenes and arrange them to demonstrate their style and versatility, being careful to find scenes that were not extensively rewritten by the story department. The tape can be impressive when different writing forms are quickly viewable as a finished product. It could look like this: two minutes of a sentimental romance, two minutes of a terror scene, two minutes of documentary, a clip from a three-camera sitcom, and finally a reality television dramatization. That would make a ten-minute demo of very interesting work. The rules for submitting scripts are somewhat rigid, and your agent should offer a tape only after contacting the story department; never send it unsolicited.

Writers more commonly use video to create an idea pitch tape. If you have a script or are pitching a treatment (a story idea written up like a synopsis that attempts to convey something of the feel of the finished product), you may want to provide a video of one or two scenes to put your concept across. Smaller multicamera video studios are excellent for this purpose, or you can use the techniques outlined in this book to produce your own.

Suppose you based the plot of your script on some particularly off-the-wall behavior of three acquaintances, and you decided to expand it into a screenplay. The initial script has been rejected by a studio because the readers couldn't visualize your concept from the strange dialogue on the pages. Now is the time for the pitch tape. Choose two or three pivotal scenes that are no more than three minutes each in length and that best demonstrate your vision. Cast the best actors you can find who need tape on themselves, and shoot your scenes.

Make sure this tape really will work for you and will say what you hoped it would. When you shoot, don't try to pretend it's broadcast video. Let your tape look like a well-produced backers' audition for Broadway. The sets and costumes should be "in the flavor of." (Think of Hamlet performed in rehearsal clothes. The actors are said to be dressed "in the flavor of.") The audio should be clear and consistent, and the camera work should look professional so it doesn't distract from your message.

When you arrange the pitch meeting, let them know you're bringing a brief video, and ask them which format they prefer: VHS or three-quarter-inch. Just before you leave for the meeting, check the tape to make sure it's been rewound.

Beginning Screenwriters

This is the first time in history that a beginning screenwriter has had the opportunity to practice the craft by writing the script and going right out to shoot it to see how it plays. Get some friends together—actors and technicians—form a little group, and go shoot your stuff.

Authors

We have chosen to include authors in this chapter for two reasons, even though they will not be directly soliciting the motion picture industry. The advice for screenwriters can also apply to authors approaching a publisher with a book or query. Many screenwriters are authors and vice versa, considerably blurring any line between the two.

These days the search for "properties" by the studios and independent producers has turned to the published novel and nonfiction book. This cuts the risk considerably for the movie maker, who can choose among subjects that have already been successful in another medium. While book publishing is expensive, it does not entail outlays of $20 to $80 million before experiencing a real return as does a major motion picture.

There is also the advantage that a publicity machine has already been set in motion, with great cost benefits for the producer of the subsequent film. As a result, even authors who have yet to consider the motion picture industry may want to do so, and could benefit from an overview of how demo or "pitch" tapes are used.

How could a tape benefit an author who's seeking a publisher? For books with historical content, the author might visit the area and make a video of the locale to reinforce interest in the subject matter. An author of a book on antique cars could visit a concours d'elegance with a video camera and let the prospective publisher see firsthand the size and intensity of the followers of that activity. A biographer could assemble a series of interviews with the book's subject to demonstrate their exceptional qualities.

As with the movie industry, it is particularly important to follow the submission requirements of the publisher. An unsolicited video will probably be rejected, but a call to the publisher for their query-letter guidelines could include a question about submitting a short video. Once you have a track record and a working relationship with story departments, publishers, and so on, submitting a video on your new subject will be much easier.

Celebrities. If your career takes you into the media spotlight, a publicist is always a good idea. Your publicist can handle your reel for you, but he or she will need the clips of your appearances that you collected before signing with the agency. For those who have yet to connect with a publicist, a celebrity reel is an important tool in the search for the right one.

Collect all your on-air material, and assemble a tape of your very best interviews. Keep in mind the purpose of the tape. Celebrity reels can be quite simple, with the opening logo of each particular show and the clips you chose from the interview. A celebrity reel could grow to a maximum of twenty minutes, but there's no harm in keeping it shorter. Eliminate other guests and any unrelated discussion by the host.

To cover jump cuts, use shots of the audience or reaction shots of the interviewer or host. An A/B roll system is not necessary for this type of tape, but if you have one available, dissolves, wipes, or effects will add a little class to the reel, as long as they're not overdone. It's best to choose one wipe effect and stay with it rather than try a different effect every time. With an A/B roll system, you can deal with the jump cuts by employing these transitions.

Celebrity reels are different from a theatrical tape and are meant to demonstrate your ability to handle interviews. They should give an idea of your personality, the material you will present, and how well you interact with interviewers. They should reveal the breadth of your career, the depth of your knowledge, and what can be expected of you when you appear on the show. The producers want to see that you're relaxed and lucid in a somewhat hectic environment.

The book tour demo. An interesting use of tape evolved as a communication between a publisher's marketing department and an author. This was a new twist on a celebrity reel for us but it served its purpose very well.

When an author is sent on a book tour, the publishing house needs to be confident that the author is going to hold up under the hot lights of television interviews. These tours are expensive to mount. The publisher wants to ensure that the publicity will boost sales.

One author came to us a few months prior to her self-help book being released. She already had some television interviews to her credit, but none dealt with her upcoming book, which she knew was going to be a hit. Her publisher felt the book would be a hit too, but being careful, didn't want to invest a lot of money in publicity. Both parties discussed the situation at great length by telephone. She wanted a big publicity tour; the publisher didn't want to take the chance. The concerns were as follows: "How do we know that you're going to come across on T.V.? How do we know you're going to stand up to all that scrutiny from the early morning program hosts and the afternoon talk show people? How do we know you're going to be charming, candid, sincere, relaxed, humorous, and knowledgeable under pressure? Oh, and by the way, sell the book, too."

She arrived at our studio with her three-quarter-inch tapes under her arm. She knew her material was strong, but feared that none of it was current enough and would work against her. Keeping these considerations in mind, we began to design the tape.

We opened with her photograph, the one that was going to be on the book jacket, and supered her name at the bottom, supported by some up beat news-style music. Next we ran a thirty-second advice-from-the-expert spot. This was her most recent appearance and the piece that most resembled her hairstyle in her photo. We used this as a bridge to the more well-known, but older, segments.

The next pieces were an interview from a morning program, and talk shows when she answered questions from the audience. Our strategy for the talk shows was to use only questions and answers that were rich with information. We made every line count. We cut out pauses, cut down questions, and condensed answers. Absolutely every line had to say something concrete. Whenever she named her previous work we were careful to leave the reference in. The final result was that you are compelled to listen to her, and you wanted to know what she was going to say next. We made sure there were no jump cuts, and no obvious or glaring edits.

When the tape was finished, it was as if the show had been shot that way. We were careful to show hosts responding positively and good cut-aways of rapt audiences. We made sure her performance reflected and reinforced everything the publishing company said they needed to reassure themselves that the author could be interviewed successfully.

We finished up with another ten-second section of an advice-from-the-expert spot that made a good tag, bookending the tape with her most recent and flattering visual material.

Total time was five minutes. She made two copies, kept one for herself, and shipped the other off to the publisher.

They were convinced and she did the book tour.

11 Business Video Resumes and Business Applications of Demo Tapes

The Business Video Resumes

It's sometimes impossible to fly halfway around the world for a job interview. Sometimes you receive multiple offers, and it is unfeasible to travel personally to every location. Certain professions require special skills that can be best demonstrated on video. The video resume can solve these problems.

Business executives, corporate public relations departments, public interest groups, and others have all developed an understanding of the importance of video press releases in presenting their points of view to the public. No longer the domain of politicians stepping up to network news cameras, this effective means of communication is available to all with the ability to use it.

Other uses of video, such as the interactive meeting, the video memo, and the video newsletter, increase the certainty that camera skills will be required of business people. Recently, Cardinal Mahony used video to explain the Catholic Church's position to workers during a recent labor dispute. His views were recorded, and the tapes were sent to the appropriate groups for viewing. Video is perfect for explaining new techniques to employees in the field or new product lines to sales representatives around the country.

A video resume is simply a job interview on tape, and all the rules of the job interview apply. Cast yourself in the role of the executive that you would be interviewing with, think of the questions that person would ask, and frame your answers. You might use some of these questions rhetorically, then continue with your answer, or you can rephrase the questions as a statement. Before you offer a video resume to a personnel director, be certain you can do a successful one. As with writing resumes, do not expect to get it the way you want it the first time. You will need to practice before you're satisfied.

One difference between video resumes and live interviews that should work to your advantage is this: In a video, you are essentially inviting the interviewer into your environment rather than vice versa. You control the elements, which means that you can choose the location and your position in it. You control what is seen on camera and what isn't. You can choose a position in the room that will present you to your best advantage. Controlling the elements as noted in Chapter 1 is a phrase producers use to keep their project in control. We discuss this concept in detail in Chapter 12. This way the stress present in a real job interview is eliminated. Be sure to make the interviewer feel welcome and comfortable. You may decide to take the interviewer on a tour of your environment, but this should not be a hand-held walking shot. Put the camera on a tripod in each new area, stopping it when you're through and moving it to the next area.

If you have examples of work to show, this can be done on the tape. Introduce and discuss each project as you present it. Photographs, blueprints, products, and awards are all acceptable and useful. It's important to smile and be pleasant and have plenty of energy. The more you appear to be speaking from the heart, the more your tape will achieve.

As with any tape, the briefer the video resume can be and still make the point, the better. Don't leave out anything important for the sake of brevity, but continuously look for shorter ways to say something or to show it.

Do not put anything on the tape that will work against you or that achieves nothing at all. The only reason you are making this tape is convince the employer that you're the person for the job. Every frame should work toward that end. If a shot does not contribute, don't use it. Generally, your video should have a very direct and personal feel to it, and remember to keep it simple.

Camera Preparation and Technique For Nonactors and Businesspeople

If you are the open, easy-going, gregarious type who is completely unfazed by stepping in front of a camera when that little red light is on, your main task will be to make sure you are in control of your material and your surroundings. It is important to have everything arranged in advance to reduce the necessity of multiple tasks. If the shoot takes longer than expected, avoid the onset of boredom. Ideas must be kept fresh to come across well, and if video problems cause the shoot to drag on, rearrange your presentation somewhat to stay spontaneous.

Most of us fall, however, somewhere between total aplomb and paralyzing self-consciousness. Actors rehearse over and over again before they go in front of anyone. They spend years learning to overcome stage fright. Everyone's heard the stories about shy actors who panic before performances; then once they hit the stage, they're fine. Self-consciousness is something that can be controlled and usually cured. The camera is not scrutinizing you. The camera is a tool, like a pen or a phone, with which you communicate your thoughts to others. It represents freedom, not constraint.

Nonactors can be even more distressed by someting about their appearance that dissatisfies them than the actors we discussed in Chapter 6. They think their nose is too big or their hair is too thin or their right side is better than their left or some such thing. If you recognize youself in this description, be aware that no one notices these things but you. Viewers of the tape will watch it for its content; they'll treat it as a communication to them by another person. Your presence is determined by your degree of relaxation, your understanding of your subject, the depth of your conviction, and an appropriate sense of humor. Being attractively dressed with a neat appearance is really the only concern. To this end, it's good to bring a friend or assistant along to check your hair and make-up and to watch that your clothes do not become disheveled. If you are wearing a jacket, be sure it is pulled down at the back so a big fold does not appear across the shoulders.

If you've decided to do a video and are, to some degree, camera-shy, get out your home video camera and begin to develop a relationship with it. You can work with it alone at home or in your office. Don't practice your planned presentation at this point, though, because you don't want to rehearse nervousness into your performance or to let the material become stale. Use almost any other subject to drain off your nerves and put yourself at ease with the camera. For example, you might record a video letter to a friend or tell stories about your childhood to show to your kids one day. This might be a good time to discuss a favorite subject. Make sure that the content is something you find pleasant and have a reason to want to do. Visualize the person you're talking to as being right inside the lens as though it were a window. Visualize your audience obtaining enjoyment from your communication. Don't bother to write a script for your practice sessions; use subject matter that you know. This will help you practice thinking in front of the camera, the same way you learned to think on your feet at the podium.

Understand that you have complete control over this camera. Unlike film, videotape is practically free. You can erase it or shoot over it. No one ever need see your performance but you. As a result, you can loosen up. Let go. Shake your hands around. Shake your shoulders. Relax. Let yourself be a little foolish. Notice that it doesn't matter. If you have a remote device to stop and start the camera, use it. If not, let the tape keep rolling, so you don't have to run back and forth. You don't need to look at everything you shoot. Forget about it until you've recorded something you feel you'd really like to see.

If you're working on overcoming self-consciousness, do your practicing alone, and run the camera yourself. You don't need the extra pressure of an audience. Eventually, the time will come when you will want to show off your new skills to someone, and you'll know when that moment has arrived. Actors rehearsing a role often refer to this moment as the *breakthrough*. It's the point where they become comfortable with their new character, and the performance begins to flow. It's a similar experience to learning a new skill like skiing. When you're skidding pigeon-toed down the kiddie slope, the last thing you need is an audience. Once you've learned the sport and your friends start calling you "Schussboomer," chances are you'll be skiing with others and not even thinking about it. When the breakthrough moment arrives, it's time to bring someone in to run the camera for you. There's no harm in letting your camera person know that you're practicing and that you plan to make mistakes. It's important to keep the exercise fun. Have some laughs, kid around. Make sure everything continues to feel unimportant.

Nerves can affect you in strange ways. We were taping an extremely witty and charming young executive. To our surprise, as she took her position in front of the camera, she seemed like a different person. When she read her lines from the cue cards, she seemed to dislike what she was saying. Had something made her angry? The tape was not going to be usable, and we had to stop. She admitted to having a sudden attack of nerves but certainly wasn't unhappy with anything. Then someone had the courage to ask, "How about smiling a little more?" (She had not been smiling at all.) "Just start your line with a smile, and end it with a smile, even if you don't feel like it." Not only did this work, but the simple exercise made her feel like smiling, and her performance improved dramatically. She became animated, interesting, and believable.

This is a trick some actors use when The Method isn't helping. Essentially, The Method—made famous by Konstantin Stanislavsky and Marlon Brando—requires a truthful actor to work from the inside out. Motivation is the key. Sometimes, however, when the wait for motivation is longer than the wait for an LA bus, something must be done. If you form your body into the shape of the emotion, as in pantomime, you can often trick your mind into believing that you're feeling that emotion. Need to feel sad? Slowly droop your chin toward your chest, curl your shoulders forward, relax your face muscles, let your mouth fall open, and pant as if you'd just been sobbing. Try it. Did it work? Okay, now throw your shoulders back and come up with a great big grin.

This leads us to another trick you can learn from actors. It's sometimes hard to go from a feeling of neutrality to the emotion you want. It's like not being warmed up. If you're before the camera feeling kind of bored and empty, and you're not interested in smiling or being animated at all, try thinking of something sad first, then quickly think of something happy. Generating emotion requires energy. Once you've got the emotional energy on hand, you can use it for a different feeling than the one that got it going.

Scripts, Teleprompters, and Cue Cards

Once you're ready to make the actual video, it's best to know your presentation so well that scripts, Teleprompters, and cue cards are unnecessary. Without the video, you would be making a personal appearance in someone's office, chatting and improvising according to the situation. Even though doing this before the camera is one-sided, the closer you can get to that feeling of interaction, the more effective you will be. In your thinking, you must anticipate others' questions, and in your presentation, you must address them logically as though they had been asked.

If you must use a script, know it backward and forward, and concentrate on communicating in a conversational manner. If you cannot avoid reading, use a Teleprompter that fits over the lens of the camera, and use cue cards only as a last resort. Cue cards are least distracting if placed to one side of the camera beside the lens, rather than above or below it. Speak your first and last line directly into the lens. Simply read the rest off the card without moving your eyes back and forth between the card and the lens. This looks worse than simply admitting that you're reading.

Setting

The setting should be appropriate to your profession and purpose. If your work is done in an office, you should tape in the office. Professionals such as architects may want to use a model of one of their designs as a background. A graphic artist could shoot near a drawing board. If these suggestions are not feasible, choose an upscale, well-structured background and be seated in a chair.

If you decide to sit behind your desk, make sure that you do not disappear behind it. Sitting on a phone book or two sometimes makes all the difference in the way the camera perceives your presence. If you would like a more casual presentation, you may feel comfortable sitting on the corner of the desk, particularly if the wall behind you is attractive, with books or diplomas displayed, for example.

A generic setting can be created with a chair placed away from a wall, a potted plant on one side, and a table lamp in the frame on the other. This helps to break up the background and to give the appearance of a comfortable atmosphere. Make sure there are no bright windows behind you. They will give everything a blue cast and make you appear to be in the dark. If there are pictures on the wall, make sure they do not cause glare in the lens. This is especially important when using a camera with pickup tubes instead of chips because the glare from a picture frame can burn the tubes. Be sure that any lights you use are not reflected by glass or metal appearing in the frame behind you.

The camera should be close to eye level whether you are sitting or standing and you should be framed from about halfway up the chest, with just a little space above the hair. If the camera operator is inexperienced, ask to lock the camera down before you do your presentation. See Chapters 12 and 14 for hints on audio, lighting, and camera operation.

Make-Up and Hair

Women can use street make-up applied a little heavier than usual to compensate for the video lights. Study your make-up on the monitor carefully to make sure you are satisfied with it. Remember that what you see is what you get. Many women hire a professional make-up artist. Make-up artists can be found by calling an advertising agency, a production house, or a professional photographer. Larger cities have make-up schools and studios. For men, a light base is allowable, but we have always found it unnecessary. A medium face powder is essential to eliminate perspiration shine for both men and women, and it should be someone's responsibility to watch for this.

Hair styles that have buns or top knots are difficult to shoot. The operator must leave too much "air" at the top of the frame in order to keep the bun in the picture, and it will look like a mistake if the bun is framed out. Hair that falls too far forward over the eyes makes you look as though you're hiding and causes odd shadows from the lights.

Product Demos

For a simple product demo on a subject with which you are familiar, writing a script is probably unnecessary. When the subject is complicated enough to require a script, it might be time to consider professional production. It is a good idea, though, to put together a shot list of everything you want to show and an outline of everything you want to say. Most art supply houses sell blank storyboard sheets, and these can sometimes be very handy. They have small rectangular boxes representing television screens—in which you draw an approximation of what the camera should see—and a place for dialogue or notes. This will help you visualize how the tape will be organized, and you'll get an idea of what the finished product will look like.

When shooting in the workplace, make sure that your background is clean and presentable. You will often find electric cords hanging down, and no one will notice how untidy they look in the shot until it's too late. If the background is messy, you can choose another or buy some photographer's background paper to cover the unpleasant area. Make sure that where you're shooting is quiet enough to get good audio. Do a test, and play it back in another room before recording the entire presentation.

The simplest way to produce your own video of an activity is to record your description of the entire process as a "talking head," then shoot the footage required to support or illustrate your words. Use a tripod whenever possible. Make sure that what you're demonstrating is clear and centered so that no unimportant background fills the right or left of the frame. Take care that nothing distracting occurs behind the camera as you are shooting, and find an angle that is interesting and pleasing to the eye. Before you leave the location, review everything you've shot and make sure all your points were made and were correct. If anything is missing, you can shoot again, out of context, and assemble it during editing.

Avoid fancy zooms for the talking head portion because it's distracting to cut in the middle of a zoom when editing. You can vary the framing, however, by reframing when you stop to think or look at your notes. The close-ups should not be too tight—always allow at least a little collar to show—and the wide shots should not get any wider than the waist.

If you're shooting a machine, let the viewer see the whole machine first, then reframe to take close-ups of important things like control panels and the workers operating the machine.

Think of the camera as a guest whom you're showing around for the first time. When you review the tape after each setup, once again make sure that you have not included any unwanted background in your wide shot before you move on.

If you own other video footage on your subject and the quality is acceptable, by all means pirate it for your new presentation. You may use interviews with other people and your own commercials or product demos that may have been made for trade fairs or conventions. News footage praising your company or product is great to include. Review all your material to see what can be used and where it can be placed in the presentation. It's important to exclude different footage taken of the same activity in the same location, however, as the different picture quality and lighting will be unacceptably distracting. You will also need to check the sound quality to make sure that it will be reasonably consistent when the different scenes are cut together.

Once you have all your footage and you know that you are satisfied with it, book an editing studio. Intercut the video of the activity and your talking head so that it flows smoothly, is logical, and maintains the viewer's interest. Make sure that neither shot goes on so long that the viewer begins to forget about the other. The activity is the most important part and will be the most visually interesting. Use your on-camera description when you are most animated and most at ease. Simply cut the places where you repeated yourself or stumbled, and insert the product footage to cover the jump cut. Finally, you can have the editing studio add music behind the presentation, titles, and other on-screen graphics or photographs. The music and effects should be appropriate to your subject and never obvious or overpowering. Edit on three-quarter-inch tape or S-VHS, and make the copies on regular VHS.

Organize your material in a chronological sequence, beginning with the drawing board and ending with the product coming off the line. It's possible to use press announcements at the very opening, and as a teaser, you may wish to cut together a short montage of what the viewer is going to see in the body of the tape. (Refer to Chapter 7.)

After the teaser, arrange your material so that the story feels like it's building in excitement and the pace is increasing. Shorter clips with more activity can be used to serve this purpose, and the background music can also help by increasing in intensity. Two-thirds of the way through, when the product is really shaping up, another montage is often effective, though it shouldn't be as staccato as the teaser. After this, move straight to the final phase.

For the ending, show customers buying the product, phone personnel taking orders, customers using the product, and outsiders praising it. You might conclude with a short credit roll or a freeze frame of the product or activity.

You might line up suppliers to praise your product on camera or collar some of your associates to discuss their department or function. Satisfied customers willing to endorse the product are especially useful. Seat your interviewee in a place other than the one you used to do your own on-camera work, and be prepared to waste some tape as you chat. Sit quite close to the side of the camera, with the lens at eye level. Ask questions that lead your subject to praise the quality of the product, the speed of delivery, the uniqueness of your relationship, the dedication of employees, and so on. The subject can look into the camera or at you, whichever is more comfortable, although looking into the camera is best. You should hesitate a beat before asking each question to leave enough room to edit out your questions. (When two people speak simultaneously, it's known as an *overlap,* and it makes the line unusable when you edit.) Interviewees will generally not overlap after your question because they take a moment to think. If this problem does occur, just ask for a slight hesitation before your subject speaks so you can cut later.

When making the tape, plan to get your message across as quickly and succinctly as possible. Avoid using too much of your own talking head; just return to it every now and then to reestablish your role as the presenter and the boss. Let the pictures do the talking wherever possible. Use the voice-over narration from your on-camera material where it is needed. Depending on the context and the complexity of the activity, a tape can last anywhere between six and eighteen minutes. If it contains educational material, the tape will have to run as long as it takes to complete the instruction.

Part II

Shooting

12 The Video Screen Test

The demo scene and on-camera audition have become known as the *screen test*. This is especially true when they are requested by a producer, casting agent, or agent. Because it's now the performer's responsibility to provide these, the universal, though ungrammatical, cry of the young actor has become, "Where can I get tape on myself?" There are four ways to do it.

1. Take an acting class that videotapes your scenes and will give you copies.
2. Have someone videotape your equity waiver play legally.
3. Hire a professional video company to shoot a demo scene.
4. Try shooting a demo scene yourself or ask a friend with a camera to do it.

The first thing to consider when you embark on this adventure is what you are going to use the tape for and under what conditions it will be viewed. Imagine yourself in the prospective agent's office. You are there with your head shots, your resume, and perhaps a tape, sitting across the desk from someone who has decided to take the time to see you. If you were in the agent's position, what kind of scene and what kind of tape would convince you to give an inexperienced actor a shot?

If the tape has a lot of clunking and banging off-camera, if your voice is hard to hear, if there is an annoying hiss on the tape, if the colors are all wrong or smeary, if the camera is so far away you can't tell who's who, if you're dressed in an outlandish costume, if you're doing vulgar things on stage, or if you're singing off key, you might as well forget it. Go to the interview, but don't show the tape. On the other hand, if the tape looks good, if there are some close-ups, if the camera is steady, if you're playing a role you might conceivably be cast in on television, if the scene is one that could actually appear on television, if the performances are believable, if you feel comfortable and confident showing this scene to a stranger, bring it in.

Shooting a Play

If you elect to get some tape on yourself by having someone shoot your equity waiver play (a play in which Actors Equity Association allows members to perform free in order to practice their craft and showcase), see if the rest of the cast will chip in. Videotaping rates vary widely, as do the results. One idea is to contact schools and colleges to see if there is a student whose work is advanced enough to do a good job and who is looking for some tape for their own demo. You might be able to make an excellent deal.

When shooting a waiver play, shoot with a single camera. Avoid multi-camera shoots, which seldom look much better, are far more expensive, and make a hornet's nest out of postproduction When you tape in the theater, the air conditioning should be off, the lighting should be as bright as possible, and the show should be miked with a couple of shotgun mikes that have been placed so as to pick up as little lighting buzz as possible. If the audio cables run a long way down an aisle, ask the camera person to make sure they are not picking up any radio stations. This problem can generally be eliminated by shortening or lengthening the cable or by using shielded XLR-style cables, which are professional cables much less susceptible to electrical and radio interference.

If your play is a comedy, you'll want to tape it with a live audience. If it's a drama, it's sometimes better to shoot a rehearsal so you don't end up with coughing on the sound track. When shooting a rehearsal, however, avoid the temptation of constantly cutting to do things over. This will cause the editing costs to go through the roof. Remember, save money by limiting your creative choices.

If your play is well rehearsed and you are happy with your performance, once through should get it on tape. You may want to arrange to shoot the play twice, once during a matinee and once during an evening performance, then choose the best performance of the two. If you are not happy with your work in this show, it's probably best not to tape it, but get a trusted, objective opinion on that.

Shooting a Demo Scene in a Professional Studio

Many small or industrial studios in larger cities will shoot single- or multicamera scenes for you. Your best deal in a studio is to shoot multicamera and have the scene switched on the fly. They generally offer one-, two-, and three-camera shoots. A three-camera shoot is preferable to a two-camera shoot and could be cheaper than shooting with one camera and editing. When you shoot single camera—film style—you do have more control over the taping; however, use our suggestions in Chapter 13 to limit editing costs. If there is a marked difference in price between the single-camera and the multicamera shoot, opt for the single camera and decide in rehearsal where the wide shots will be, where the close-ups will be, and where the camera moves will be. The less movement you block into the scene, the fewer times you will have to cut due to focus problems.

Select the studio with care, using the criteria outlined in the following list, and study the costs carefully. The best way to find out about a studio is to check with people who have already used it to discuss their experiences. Ask previous clients the following:

- Was the actual price relatively the same as the quoted price?
- Were the people in the studio easy to get along with or was there friction?
- Did the length of time the shoot actually took match the time the studio told you it would?
- If the time estimate was far off, was it due to your underestimation or the studio's?
- Did studio personnel try to convince you to pay for extra things, such as special effects, extra editing, or an elaborate set? Were there any hidden charges you didn't know about until you got the bill?
- Was the studio fairly clean and fairly comfortable?
- Was it necessary to cut for nonartistic problems, such as phones ringing, people knocking on the door, or trucks rumbling on the street?
- Was it necessary to cut because of operator inexperience or machine failure?
- Were you expected to pay for the time the studio was down?
- Were you allowed to provide your own tape, and were there any extra charges for doing so?
- Who owns the shooting tapes at the end of the shoot, and who owns the master? If the studio owns the master, you must have all of your copies done there, whether the rates are competitive or not.

If you have checked with friends and not found a studio you're happy with, go through the Yellow Pages and the acting and production guides, such as the *Working Actor's Guide, 411*, or any of the others. You will find them in specialty bookstores or in the library. Pick studios that appear to be in your price range, and start phoning. Listen to the voices on the other end. Do they sound happy or harried? Are they helpful or noncommunicative? Are they being evasive and vague? Editing houses must sometimes be vague at first until they have enough information about your project to make a cost estimate. If that's the case, ask studio personnel what it is they need to know in order to be more specific.

Cost estimates will vary widely, and as with all video, the cost is not always a reflection of the quality. Your defense is to study the studio's sample tapes and to make sure you feel comfortable with the people with whom you'll be working. For a multicamera shoot, quotes will range anywhere from $75 an hour and up. Unusually low rates can sometimes signal problems with the studio, so don't leap at them. A good price range for a three-camera shoot of a demo scene in 1991 was $100 to $200 an hour. Don't imagine you can get the scene done in one hour; it just won't happen. If you follow the guidelines in Chapter 13 for choosing and rehearsing a scene, though, you should be able to get it shot in two.

If you get together with friends and book a block of time at a reduced rate and you decide to take the leap into editing and postproduction, avoid doing it all on the same day. The shoot will be enough. Everyone's creativity will be virtually burned out afterward, and the editing session will be slow and lackluster, if not downright crabby. Go back in the morning with some coffee, and everything will feel like fun again.

Selecting a Video Person to Do the Taping

A skilled and caring camera operator will produce the best results, and this is the person you should be shopping for. Use the following guidelines when picking any studio or campany to do your video for you.

DEMO TIP: What you see is what you'll get.

If you decide to find someone to shoot your scene or waiver play for you, select that person carefully to ensure that you get the desired results. Whether it's an acting class, a professional studio, or a friend who will handle your taping, assess their previous work carefully before you jump in. When you look at their samples, find out if your scene will be shot under the same conditions. Ask the following questions:

- Will you use the same camera?
- Will you mike the scene in the same way?
- How much did the final product cost?
- How long did the scene take to shoot?
- Was editing involved?
- How long did the editing take?
- Will the same camera person be used?
- Was lighting involved?
- Will the same lighting person be used?

When looking at sample tapes, watch carefully for the following problems:

- Is the picture grainy and smeary?
- Does the scene look dark and the focus soft?
- Are the faces green or magenta or bright orange?
- Does the audio sound rumbly, garbled, or too bass?
- Does loud hissing on the tape make the voices sound distant and difficult to hear?
- Is the air conditioner the loudest thing on the sound track?
- Do the voices suddenly cut in and out?
- Is the scene shot entirely as a long shot?
- Is it shot entirely as a close-up?
- Is there a lot of unnecessary zooming around?
- Is the camera wobbly or unsteady?
- Is the background distracting?
- Is the lighting bright in one area and dark in another?
- Do the actors look good in one segment and blue in another?
- Is there clunking or banging or talking going on off-camera?

If you run into these conditions, the result will not be a tape you'll want to show to an agent. If you have to show it because you promised to, you will not be proud of it. Remember to choose your video person as carefully as you would your photographer. Think about head shots for a moment. You can be the most accomplished actor in the world, but if your head shot is poor, casting agents will pass on you. It doesn't matter how close you came to getting the job, if your head shot failed you, you're unemployed. There are only two kinds of jobs in show business: those you got and those you didn't.

Shooting Your Demo Scene Yourself

Suppose you've researched your options thoroughly and you've decided that the way you will have the most control over your product, given your budget, is to go out and shoot your demo scene yourself. Ask yourself, "What do I want to do and what's the best way to get it done?"

One suggestion is to get together with others from your acting group, theater club, workshop, or scene study class, rent equipment for the weekend, and shoot everyone's scenes at once, sharing the cost. Once the scenes have been shot, book an editing studio to do whatever cleanup editing is needed.

It's now time to consider preproduction.

DEMO TIP: Preproduction is the key to success.

Before recording anything on tape, think about the following:

- what you want to do and why you want to do it
- how much money you can spend on it
- what the return will be (in terms of fulfilling your objectives)
- who you want to work with
- how much time you'll need
- how much time you'll have
- your skills and limitations
- the skills and limitations of others
- the location
- getting to the location
- what control you have over the location
- your audience
- what the audience will think about your product

Write your thinking down.

DEMO TIP: Keep it simple.

In the somewhat gruff world of parachuting, instructors say it like this: "KISS." They added the little word *stupid* to the phrase so that when their students are hurtling toward the Earth at 120 miles per hour, they will have a neat acronym to hang onto in case this is their moment for a brush with another favorite phrase of theirs: "It almost always works." In the slightly safer world of video production, when hurtling toward overtime at about the same velocity, that acronym comes in handy.

Let's assume you decided to do a group taping session but didn't have time for preproduction. "How complicated can it be?" your friend reassuringly asks.

DEMO TIP: Never solve a problem when you can avoid it.

Shoot day arrives and you're all set up in the room you rented. Tape is ready to roll. The actors of the first scene are silently clutching their emotional memories, poised and listening for that most magical of all words, "*Action!*" You whisper to your small crew, "Okay, quiet. Let's settle, everybody. This is picture!" The air is electrified with the expectation of transcendental video. This is the precise moment when an enormous rattling engine cranks up right over your head. The air conditioning, it turns out, is not under your control.

What can you do? You might try to solve the problem by finding the building's janitor and engaging in an act of melodramatic pleading. Then you remember that this is the man you met when you propped open the alley door to load in the stuff and how he explained to you that he is a man who believes his life would be heaven if only there were no actors in the world, especially you. Okay, what else can you do? Maybe you could solve the problem by sneaking up on the roof with a giant crowbar. Where can you get a crowbar?

The best solution is easy: avoid the problem. Move to a room without air conditioning. Keep it simple, you see? Especially under stress.

The time to really avoid the problem is during preproduction, when you should attempt to get *everything* under your control. Visit the room. Find out, for instance, who controls the air conditioning. If there's the tiniest hint of doubt in your mind, find another room.

The best time to discover potential problems is during your location scout. Try to meet everyone who might be there on the day. Shake hands with the janitor and ask where you should load in. Ask if you can prop open a door as long as someone stands guard. Would the janitor like to be there personally? How does the air conditioning work? Can you get it turned off when you shoot? And so on. The janitor might be so tickled with your attitude that he or she will help you carry in the gear!

We love looking for ways to avoid solving problems. Here's one that happens all the time: The background wall is white. You discover that this condition makes your white actor black and your black actor nearly invisible. To solve the problem, you shut down and go rent every light, scrim, flag, barn door, and mounting bracket ever invented. But then, before you can yell, "We're back!" you have to spend four more hours trying to figure out how to keep all that light off the back wall and still get enough on the actors. All this time, the wall behind the camera was brick.

Turn the camera around, repo your actors, light them up with the three original lights you rented, and shoot your scene. The light won't bounce off the brick, and the performers will look just fine. You didn't have to solve the problem because you avoided having a problem to solve.

The Crew Director

Before outlining the recipe for a simple and economical demo scene, we need to talk for a minute about the person who will be in charge of the taping. The advice in this section applies whether you're taping a bunch of scenes or just one.

Whether you shoot your scenes in a group or do it individually, someone must serve as crew director. Note the word *serve*. When you select a member of your group to serve in this capacity or if you are selected by your group to serve in this capacity, it can be very exhilarating and a lot of fun, but it can also lead to murder. Here are a few pointers to make this a happy experience for everyone.

DEMO TIP: Do not show up in jodhpurs and a beret.

The title *crew director* describes the limit of the function. Refrain from directing the acting unless you are the established acting coach or leader of the group, and never play *auteur*. Unless the acting coach is on the set, the acting is the actor's job. Your job as crew director is to get whatever the actor is doing on tape as nicely and as unobtrusively as possible. If you make sure everyone has this straight before shooting starts, everyone will have a pleasant day. Help the actors into the best blocking positions for the camera, and let them see what you mean by showing them the monitor as you block.

DEMO TIP: Paste a broad grin on your face the minute you walk into the room, and don't wipe it off until you wrap for the night.

DEMO TIP: Never say, "Yes, but . . . ." Always say, "Yes, and . . . ."

Any negative words are a turnoff to creativity, and they introduce doubt, the enemy of performance. Acting is gossamer. It happens in the air between people. Never say, "No." Never say, "Stop." And when you do say, "Cut," say it matter of factly, then immediately explain why you said it in positive terms. Avoid saying, "You walked out of the light"; say, "We need to help you stay in the light." Avoid saying, "You said the line wrong"; say, "I'm wondering if that was the right line there." Tact, encouragement, and a positive attitude are just part of the job.

Remember, it's not an airplane; it can't crash. Frustration is your enemy, and Zen is your friend. Part of the excitement and part of the reward of being a crew director is avoiding problems before they arise and solving those you absolutely must. Keep everyone informed of everything that's going on. This is not the time to score points for arcane knowledge. This is the time to help everyone get the most dynamite scene they can.

For individuals or groups that can afford it, it is possible to hire professionals for certain jobs like the camera operator. If you do, it is the crew director's responsibility to make sure that this new person is there to do no more than run the camera, help with the lights, and shoot what the actors have rehearsed. Once you start hiring people for your shoot, you'll probably be better off booking a studio and letting the studio handle the whole thing. That way, you walk in, they shoot your scene, and you go home with your tape.

If you decide to stay with the do-it-yourself route, the following guidelines will help you come up with the best possible product for the lowest possible price.

Audio

DEMO TIP: If you didn't get the audio, you didn't get the scene.

People are generally so intent on the video that the audio is all but forgotten. It even happens on a real set while shooting a real movie. Everyone's looking at the camera or the actors, and no one's looking at the soundman, who's sitting around the corner, way back in the bushes, under his headset, behind his little cart, where he can hear.

Take 16 has just ended. Everyone is holding their breath. The actors look up anxiously, hoping they've said all their words. The continuity supervisor stands frozen in time, right at the moment that she clicked her stopwatch. The director, who has just yelled "Cut," is staring at his toes and feverishly thinking through everything that seemed finally, thank goodness, to have gone just right. The assistant director is still holding the traffic; she stares intently at the director, waiting for a sign.

Finally, seconds (which seem like ages) later, the director mumbles "Okay for picture?" The camera crew nods. "Okay for actors?" The dialogue coach nods. "Continuity? Did they say everything?" The continuity supervisor nods. "How were we for time?" Continuity whispers, "One and a half seconds over. You can live with it." The director's almost imperceptible look to the AD gives her the signal, and the AD then bellows, "Okay, everybody. We got it! Wrong set! We're losing the light! Let's hustle!" At that moment, screaming out of the shadows like an enraged bear flies the sound recordist, flailing his arms and yelling, "Not okay for sound! Not okay for sound! Not okay for sound!" It happens because the sound recordist has to sit so far away from everyone to hear, and everyone's all wrapped up in making the picture.

Selecting an appropriate microphone.

DEMO TIP: The closer the mike is to the actor, the better the sound that you'll get.

Wireless lavaliers attached to the actor, approximately at the breastbone, are ideal for an acting scene. If the actor looks constantly to one side during the scene, favor that side with the mike. You can usually rent professional mikes from a rental house for a very reasonable rate, or you can buy perfectly adequate ones at a discount video store or at Radio Shack. Azden makes an excellent selection.

If you are in an area where there is electronic interference or heavy fluorescent lighting, use a wired lavalier (lav) and hide the wire. To hide a mike and keep it from rustling against the clothes, double up a little Gaffer tape (or duct tape), and tape the mike to the fabric. If there are two pieces of fabric—for example, a shirt and a tie—tape it to both. If you're shooting an on-camera audition, there's no need to hide the mike. Everyone expects to see it, and you will avoid having to cut because the mike is rustling against the clothes.

Another excellent mike that can be used to pick up many sound sources at once is a pressure zone microphone (PZM). This mike supposedly was first used to record the banquet scene during the final episode of M.A.S.H., but now you can purchase one of these superior mikes in places like Radio Shack for around $35. An advantage is that they are completely flat and easy to hide in the scenery. They use wooden surfaces as a sounding board, so they are great to place on a table but should not be set on a stage floor where delicate footsteps will sound like someone is crashing through the jungle.

As discussed in earlier chapters, it is usually impossible to fix audio problems in postproduction. If clothes are rustling against the mike, this exceptionally distracting noise will be mixed with your lines for good.

Checking the audio.

DEMO TIP: Never check your audio in the room where you recorded it.

Think about the sound man who sat so far away that everyone forgot about him. The human ear adjusts to the sounds of its environment. It screens out irrelevant noises and hears only what it wants to hear. If you're in a room where an air conditioner is humming away, unless it's very apparent your ear will ignore it. If you play the tape back in the same room, your ear will screen out that same sound on the tape as well. When you get to the editing studio, though, you will notice it. The operator may suggest equalizing it, which will help only a little bit.

There are all kinds of other noises to be alert for, and it's something you need to practice. Take airplanes, for instance. First they build a studio, then they build an airport to help everyone get to the studio. At least, we think that's how it works. If the scene supposedly takes place in an office tower, watch out for neighborhood dogs barking. Listen for people whispering during the take. Did the mikes pick it up? Did anyone drop anything? If you plan to edit, this gets tougher. A siren going by will be a nuisance in post because you will hear it on one close-up but not on the other. To avoid this, cut and go again once the siren has passed.

Keep the camera on manual rather than auto volume control, and make sure the meters never "pin," recording so loud that the audio distorts. Many auto-level features on cameras continue to push up volume levels until they pick up a sound. This means that unacceptable hissing will rise up during every pause in the audio—during the big kiss, for example. Set the levels for the loudest part of the scene, making sure the quietest part is audible, and unless something changes, leave it.

Lighting

Tools required. Go to your local drug store to purchase what you don't already have of the following:

- a roll of duct tape, which is cheaper than Gaffer tape
- a pair of cheap work gloves for handling hot lights
- a few sheets of white construction cardboard to bounce light
- a pair of scissors
- transparent tape
- at least two long and two short extension cords with three-prong plugs
- two breaker-equipped electric plug strips
- a bag of clothespins (known jokingly by gaffers and grips as "C47s")

From a camera rental or sales establishment you will need the following:

- three 32K video lights, usually with 600-watt lamps and barn doors
- a couple of sheets of "spun" Dacron, which softens light

Look under "Video Equipment Rentals" in the Yellow Pages. The rental houses that cater to the industry often have cheaper rates for better lights than do video stores and camera stores that rent to consumers. You should be able to rent a well-equipped light kit for $35 to $50 a day. If you rent it for the weekend, many houses will charge a single day's rental for both days. Ask what comes with the light kit; much of what you would buy at a drug store may be included. Make sure there are a couple of spare lamps in case one burns out. Lowel-Light Manufacturing Inc. is one of several companies that make excellent video lights. They are light, compact, and easily manipulated.

White balance and color temperature. The camera decides which colors are which by comparing them to white. If the color of the lights under which the camera "sees" white changes, it will get the colors wrong. When this happens, the white balance is said to be off. The color of light is measured by determining the color temperature of the light.

DEMO TIP: Never mix 32K light and 45K light.

Here's a little story out of our third-grade electronics engineering reader. We don't know if it really happened this way, but if it didn't, it should have. Once upon a time, a man named Kelvin (K. for short) wondered why the sky was blue, the grass was green, and how to tell the difference in between. He meant the exact difference—and exactly the same exact difference, every time.

One day, he was hanging around a blacksmith's shop as the blacksmith was heating up a piece of iron. He noticed that the color of the metal changed as it got hotter, from red to yellow to blue to white. "White Hot!" he yelled. He ran off as quickly as he could and set up his own experiment so that he could measure these changes scientifically.

When the temperature of this piece of iron reached 3200 degrees, he had a nice warm indoor color to light up all the different regular colors of the things in his room. By the time it reached 4500 degrees he was seeing things in the bluish light of nice bright sunshine. He named the first type of light *3200K* and the other type *4500K*, and his scale went all the way up to 6500K for glaring white sunlight. At this point, these names had nothing to do with temperature any more and referred only to the color of light. We perceive reddish light as warm and bluish light as cool. Thus, the lower the color temperature, the warmer the light, and the higher the color temperature, the cooler the light. So 32K, which is the standard for video lights, is a nice warm light, and 45K, a standard for shooting outdoors, is bluish.

DEMO TIP: Always white balance the camera, and always use the light you will be shooting in to do it.

On a professional camera, the filter is set to 32K before it is white balanced under 32K light. Outdoors, the filter is set to 45K before the camera can be properly white balanced in 45K light. Some cameras have a filter for 65K light as well, and this filter should be used in very bright sunlight and on white sand or snow.

Consumer and one-chip cameras simplify the process of white balancing by just asking you to point the camera at a white card or object that fills the screen and to push the white balance button. Some cameras have a white translucent lens cap, and to white balance, you can leave the lens covered and point the camera in the direction of your scene, then push the button. If there is not enough light, the camera cannot be white balanced. If that happens, you must set the camera to a preset white balance or use the auto-white setting, if there is one.

What happens if you have lit your scene with 32K video lights, but at one point one of the actors walks over to a window? Because the camera has been white balanced for the 32K artificial lighting, when the actor gets to the window he and everything around him will be blue, and the result will be distracting. To solve this problem, you can do the following:

- Buy a sheet of very expensive amber gel made to convert 45K to 32K, and cover the window with it.
- Buy some very expensive sheets of blue gel made to convert 32K to 45K, gel the lights with it, and shoot the whole scene in 45K light.
- Put opaque curtains or a Venetian blind over the window, and have the actor peek out.
- Use a fake window as a set piece.
- Have the actor walk off camera to the window.
- Have the actor simply allude to the window and stay put.
- Cut the dialogue about the window.

Let's review a few demo tips that are especially important in lighting:

- Never solve a problem when you can avoid it.
- Keep it simple.
- What you see on the monitor is what you get.
- Never mix 32K and 45K light.
- Always white balance the camera, and always use the light you will be shooting in to do it.

The three-point lighting setup. There are three positions for the lights used in a three-point lighting setup, and each one is named after its function: the key light, the back light, and the fill light.

The *key light* is the main light source, and it should be set just off to the side of the camera, somewhat higher than the actor's eyes and high enough so that distracting shadows do not appear on the background unless they are cast off-camera to one side or the other. It's acceptable to have one normal-appearing shadow as long as it is not overpowering, but you should never see a double shadow on the monitor.

When setting the key light, one of your objectives is to keep the shadow cast by the actor's nose as small as possible. The closer you set the key to the camera, the less shadow you will have. However, it is a nice effect to keep one side of the actor's face slightly brighter than the other to produce a three-dimensional modeled look. The closer the light is moved to the camera, the more even the lighting will become on both sides of the face.

The *back light* is the second most important light. Some news reporters who go on location interviews carry only two lights: the key light and the back light. The back light does interesting and artistic things, and its effect can be downright exciting. For the pretty star and the handsome leading man, it puts a glittering, dancing light in their hair, and enough back light can create a halo.

It does other things as well. Video is a two-dimensional medium. Everything on the screen is as flat as a pancake. To create the illusion of depth, we use the back light. Besides lovely, glittering, shining light in the hair, it makes a little edge of light on the shoulders. This separates the actor from the background. The back light lights and defines other things in the background, so we get a sense that the performers are some distance from the back wall. We also get an idea of the size of the room they're supposed to be in.

The tricky part of the back light is making sure it doesn't cast a double shadow on the front of the subject's chest. You will need to adjust it until the effect of this shadow is minimized. If you can't hang the light, a second problem is to hide the stand. This can be done by hiding it behind a plant, a shoji screen, or another object or by placing the stand out of the frame altogether at an angle to the actors.

The *fill light* is much softer than the key, and it is used to light areas on the opposite side of the performer. When two actors are performing together, one actor's key is the other actor's fill. When a key functions as a fill, it will automatically be softer because it's farther away.

Setting up the lights.

DEMO TIP: All lighting is compromise.

If you get one thing perfect, something else will be off. Adjust all the lights until there is a balance.

DEMO TIP: Lighting is the part of the production that takes the most time at the moment when there is no time left to be taken.

Tell everyone, including yourself, that you will need time to set the lights so that everyone will look great. This is one of the moments when you congratulate yourself on having had the foresight to bring some food to the set.

DEMO TIP: You don't need any more light than you need.

Make sure your monitor is adjusted correctly, and watch the monitor intently. When the actors and the set look good and no area is too bright or too dark, you've got your lighting set up properly. If you're lucky enough to have a waveform monitor, make sure that nothing goes over 100 IRE. If you see any bright white spots—referred to as *blooming* in Part III—you are probably over 100 IRE and should take the appropriate light down until the spot looks normal. To take a light down, you can adjust it to the "flood" setting or add spun Dacron or diffusion material. If you have to take it down too far, try moving the light instead. When you get a hot spot—blooming—caused by perspiration shine on an artist's face, it can often be solved in a minute with powder instead of the time it takes to reposition the lights.

An actor's scene is generally best served by soft, even lighting, and this can be accomplished most easily by adding diffusion to the lighting instruments. The light kit you rented will probably contain diffusion scrims, screens, and filters. These can be used to soften the light, and if you need more softness or a different feel to the soft light, cut up some pieces of spun Dacron and attach them to the barn doors with clothespins.

The best way to soften and spread the lighting is to bounce it. When you rent a light kit, ask for one that includes umbrellas. Fit the umbrella to its mount on the lighting instrument itself, and turn the light around so it faces away from the subject. The light will then bounce from the umbrella onto the set. The back light is usually scrimmed rather than bounced if it needs to be softer. Softening the back light too much will defeat its purpose.

Another way to bounce light is to use white cards. Tape them to a wall or a stand—almost anything will do—so that when you bounce the light off the card it will reach the right area of the set and serve its purpose as a key or a fill.

White cards are also used to add fill light to specific areas. You may want to balance the key light reaching an actor's face by bouncing some from below, particularly if a lot of light is hitting the actor's forehead. Use a white card for this. If there is an area of the set that is a little too dark or if some business occurs in an underlit area, a white card will do the trick. You can get quite creative with things that bounce light. If you need to bring up the light a bit on an actor in a restaurant scene, put a white plate in front of her. The light on her face will increase a full stop.

When using lights, be sure that you don't overload the circuits, avoid plugging into two-prong outlets, and know where the breaker panel or fuse box is. Try to plug lights into different circuits by running a long extension cord down a hall for some of your power, and try to separate lights from video plug-ins if you can.

Lighting outdoors is similar in concept. The sun is the key light, and bounce cards or reflectors can provide back and fill. Positioning is important. Select your shots according to the time of day and the movement of the sun. Contrast is a problem outdoors, so avoid shooting under trees and so on unless your shot has been well thought out and you have plenty of ability to reflect and bounce light. Overcast days provide the softest light, and late afternoon—around 4 P.M.—is known as the *golden hour* because of the beautiful, romantic light it provides.

Lighting is a career, and there are many master artists who sculpt with light. For actors shooting their own scenes, simplicity is the key to success. Get things as even as you can, remember that what you see is what you get, watch the monitor carefully, and give yourself enough time to get it right.

Shooting On-Camera Auditions and Static Monologues

It is important in on-camera auditions to get the very best picture you possibly can because the more aspects you have in your favor, the better your chance of being cast. However, casting departments have been using consumer cameras for years in these auditions, and if you're on a strict budget or short of time and you have a late-model video camera at home that generates a good picture, you can save the expense of booking a studio.

All of the rules for shooting two-person scenes apply, but the elements are much easier to control.

Positioning Talent

Place a stool several feet from the back wall if you plan to sit. In a shot where you're going to be standing, tape some marks on the floor for your toes so you'll always be in focus. If the only wall you have is white or some other bright color, buy a roll of background paper at any photography store. Choose a soft color such as medium blue (chroma key blue), medium brown, or medium gray, and tape this on the wall behind you so it fills the shot. The background should make your face the brightest part of the picture.

Lighting

If you have three video lights, set them up as described previously with the key light near the camera. The camera should be far enough back so that you can be framed in a medium close-up when the zoom lens is set at the midpoint between all the way out and all the way in. It is very important in this instance to bounce the light for a pleasing shot, so use a photographer's umbrella or a bounce card (white cardboard construction paper). Face the key and fill lights away from you, and arrange the umbrella or cards so the light is reflected onto your position. The back light should have a diffusion screen or a piece of spun Dacron in front of it, and it should be positioned off to one side at the back, the same distance away as the key and just out of the shot.

If you have only two video lights, use one as the key, one as the back light, and forget about the fill. If the only light you have is the one that mounts on the camera, remove it from the camera and place it where the key light would be positioned. Because lights of this type are of lower wattage, move the light close enough to you to give a pleasing effect and aim the light directly without bouncing it. If the light has a diffusion filter, use that.

Miking

When planning the audio, remember that the closer the mike is to you, the better the sound that you will get. The best mike for this job is a lavalier, and you can buy one that will work just fine for about $20 at electronics stores. The next best mike is a shotgun mike placed on a stand in front of you and just below frame.

If you must use a camera mike that cannot be detached, the sound you will get will probably be acceptable because the camera is much closer to you than in a two-person scene. Try to work in the quietest environment available. When choosing the best place to shoot, listen for traffic noise or street rumble, air conditioners, refrigerators, off-camera voices, telephones, and so on before you decide.

Working with Text

Always memorize the dialogue or copy. Go out of your way to make this possible. Get the sides (the pages of script with your lines on them) as early as you can. Learn the secrets of quick memorization. Books on improving your memory sometimes contain suggestions that are very effective for actors. If you simply don't have time to memorize the dialogue, at least learn the first and last lines.

If you're talking to the camera, say those two lines directly into the lens, and read the cue card for the rest.

The Performance

In an on-camera audition, begin by looking directly into the lens with a happy smile, and introduce yourself. If you have an agent, you may want to add the agent's name, but that's optional. Take a very short pause to compose yourself, and begin. If you have been asked to put down more than one take, name each take, but slate your name only once at the beginning. At the end, take another very brief pause before leaving the frame. It's fine to say, "Thanks" or "Thanks for watching," and that's preferable to saying, "Cut," "Scene," or "End." Avoid frowning or making a cutting motion with your hand.

13 Making Your Demo Scene a Success

Choosing an Appropriate Scene

DEMO TIP: You can do anything you want.

Preproduction also includes deciding what it is you want to shoot. You are in control of your material. You can do anything you want. If you can't find a single scene that suits you, write one.

All right then, why not rent an old railroad car, haul it out to the desert, wrap up in some sheets, stride around on top of it and do that magnificent scene from David Lean's *Lawrence of Arabia?* If you can do anything you want, that also means you can choose what you'd be better off *not* doing. Apply the keep-it-simple rule, and you'll think of a few reasons why you shouldn't tackle *Lawrence*.

By avoiding the blinding sky, the white hot sand, the huge establishing shot (which says nothing about your acting), the possibility that the location is an abandoned nuclear test range, the deafening wind (which necessitates post-syncing the whole thing once you get home), the white towel across the lower half of your face (which makes it impossible to see what you look like), the limited opportunity for close-ups, the danger of walking around in a flowing sheet on a narrow roof with stubby little pipes sticking out of it, and the seven days of round-the-clock setup involved for a three-minute scene, you will have also avoided comparing your performance to that of one of the world's greatest actors in one of the most memorable movies of all time. Who needs that?

Simplicity

DEMO TIP: Pick your scene not only for content but also for simplicity.

You want a scene that is going to make you look great. You want a scene that is going to get you an agent. You want the voice on the other end of the phone to say, "We've decided to sign you!" That's all you want.

Don't pick a scene where you have to eat a lot of food. It doesn't say anything about your acting, and if food falls out of your mouth, you'll have to do another take. If you have to do a pickup, there'll be a big discussion about whether you had already taken a bite out of the banana or not. It's a waste of time. If you shoot the pickup and the banana is suddenly whole again, the laugh you'll get when it's cut together will not be the one you want.

Don't choose a scene where you have to run all over the place. The camera operator will never be able to keep you in focus, and you won't get any close-ups.

Don't pick a scene where you must spend half your time in a screaming tirade. Keep in mind that casting agents want to see your entire range, not just the top of it, and with yelling like that the microphone will probably distort so badly, your lines will be unintelligible. The microphones used in video today are very good, but their top end is still limited. If there's distortion, you won't be able to use the tape.

Don't pick a scene where you're all bandaged up, playing a character you'd never be cast for, or doing something particularly vulgar or repulsive, and don't pick a scene that requires a noisy location.

Control

DEMO TIP: Choose a scene that allows you to control its elements.

In addition to choosing a scene that is simple, it is vital to choose one that will allow you to control all of its various elements, including the light source, the size of the set, your movement, the camera's movement, and the environment.

Control your light source. White balance is critical in video, particularly in regard to outdoor and indoor light. If you choose a scene that requires you to move from outdoors to indoors, the color temperature will change. If your camera is set for indoor lighting (32K), you will look blue in outdoor lighting because the color temperature outdoors is 45K. Pick a scene that keeps you in one type of light or the other.

DEMO TIP: If the set feels too close for comfort, it's probably just right.

Control the size of your set. Television exaggerates distance. You and your partner should feel a little too close together, and the set should feel a little too cramped. If this is the case, it will look just right to the viewer, make the job much easier for the camera person, and greatly simplify lighting and sound. Remember that you are shooting the demo scene to showcase your acting. Although the set should be as real as you can make it, the simpler it is, the less attention it will draw. The set should make your acting easier by allowing the viewer to believe you are where you say you are. The easier you can make this on yourself, the more time you'll have to concentrate on your performance. When you choose your scene, think about how wide a set you'll need, and if the scene you really like takes place in an airplane hanger, find a way to isolate it in a small area around the coffee maker, for example, with pictures of big aircraft taped to the wall.

Control your movement. The best idea is to make an entrance in a wide shot (a full body shot), so viewers can see what you look like. After that, stay put as much as possible. The remainder of the scene can be done in loose close-ups (sometimes referred to as *bust shots*), which include your head and shoulders. Keep the bottom of the frame approximately chest level. Push-ins can be added for intensity. A push-in is a slow zoom in. It can sometimes be so slow that it's unnoticeable. It requires practice, however, and the camera person must be experienced enough to handle it.

Control the camera movement. You and your camera person can decide in preproduction and in rehearsal on the day of the shoot what camera moves will best complement your performance. If camera moves become obvious, the viewer will be distracted from your performance.

A camera can do this in two primary ways. Even when the camera is mounted on a tripod, an inexperienced camera person can cause it to wobble around and make jerky moves. The result is very distracting. If this problem occurs, the solution is to lock the camera down. Set up and focus your shot, roll tape, then keep all hands off the camera.

The second way a camera can cause distraction is through excessive zooming. The zoom lens wasn't always around; at one time, this type of effect could only be accomplished by wheeling the camera in on a dolly. Dollies are in regular use today, of course. They cannot be replaced by a zoom lens because the latter distorts perspective while a push-in with a dolly is true to what the human eye perceives. With the dolly method, at least three people have to sit on a device that looks like a railroad handcar. One runs the camera, one pulls focus, and one pushes the whole thing along. If anyone hiccups, they have to do it all over again. This system keeps excessive push-ins to a minimum.

Camera people, like editors, are extremely conscientious, and it's easy for them to feel like they're not doing their job if the frame isn't constantly changing. It's also easy for camera people to get bored with the shot early on because they are not concerned with the content of the scene. Their job is to watch the frame and the focus, and pretty soon they just can't resist pushing in a little tighter. Every now and then reassure the poor blighters that they're doing great, and they'll calm down.

To control the camera, a fluid-head tripod is of paramount importance, particularly for less experienced operators. Rent the best tripod you can afford. O'Connor 50s or small Sachlers are perfect for this job. The fluid-head tripod does two things especially well: It eliminates the jerk at the start of a camera move, and it eliminates the camera shake that can take place with cheap mechanical tripods. Those faults will make your product look amateurish even if your performance is as smooth as glass. Smaller fluid-head tripods that are made specifically for small video cameras are now reaching the market. If you plan to do a lot of shooting, you might look into one of those. Another type of tripod, which is sold in consumer stores, is the fluid-effect tripod. We'll leave it up to your hands-on judgment. Our feeling is that it's either fluid or it isn't.

The least controllable camera work of all is the hand-held look. If this is the feeling you want in your scene, get the most experienced camera person you can find, and rehearse the scene very carefully. Don't choose a scene where the operator has to stumble across rocks at a beach, walk backward upstairs, run down a railroad track, or play in traffic.

DEMO TIP: Always be in control of your environment.

Control your environment. Working actors are trained to perform in an imaginary environment. That is, they are paid to imagine things, and none of it is real. Bottles and windowpanes are made of sugar. Bullets are blanks. Chairs are constructed of breakaway balsa wood. Boulders are made of Styrofoam. Actors get used to that. Here's what is real. Traffic is real. Money is real. The edges of rooftops are real. Real guns are real guns, and so on.

The point is that everyone can get carried away with the imaginary environment and lose track of the real one. Actors depend on their director and crew to keep them from stumbling off the set and into something that can hurt them. The crew director or other assistant must not get so caught up in the "action" that they lose sight of the reality of their surroundings and fail to protect the actor.

It's possible in all the heady excitement to feel a touch immortal. On a stage, the boundaries between the imaginary and the real are quite clear. On a set, they can become exceptionally blurred. Even if what you are doing on camera is only a demo scene, it can feel, at that moment, like the most important thing in life. Be sure you are always aware of what is real and what isn't.

Suggestions for Simplifying a Successful Demo Scene

Choose a three-minute scene. Choose a character that you would really be cast to play. Choose a scene that allows you an emotional range.

Make sure the scene you choose is suitable to both male and female agents, keeping in mind that there are more women than men in the business. Scenes that portray a lecherous, grasping agent or producer unabashedly maneuvering a witless young starlet toward a casting couch might well be avoided.

Keep eating or drinking in the scene to a minimum. Be careful about clinking cups on saucers and glasses on tables. Cover the saucer with a paper napkin, and always put a tablecloth on the table.

Block your moves in an arc, side to side, loosely around the camera. If the blocking is upstage and downstage—that is, toward and away from the camera—you'll have focus problems and you'll have to cut unnecessarily.

Decide on a simple set and background. A brick background is excellent. A white background should never be used, and wood paneling reflects lights annoyingly. Break up the flatness of a back wall, curtain, or cyc by arranging some plants in the background. Add a piece of lattice or anything else that can provide interest for the eye without being distracting. (Make sure the plant is off to one side and doesn't appear to be growing out of your head.) The secret to both staging and lighting is to keep your set confined. If you can stage the scene in a corner rather than across a flat wall, you will add three-dimensional depth to the set. If you choose a corner, make sure you can still use back lighting effectively. You may need to hang the lights or to hide the stands. If the corner is going to be a problem, forget it and move back to the wall. Confine your set in another way.

Use the three-point lighting plot outlined in Chapter 12.

Use lavalier mikes, either wireless or wired. If all you have is an ordinary mike, use a boom to get as close to the actors as you can. Your third option is to position shotgun mikes on stands as close as you can to the actors. Never use the mike attached to the camera; it's too far away. You'll get so much room noise that the scene will be hard to hear and will have echoes, and the overall impression will be amateurish. Camera mikes are improving all the time, but it's best not to take the chance unless you're already familiar with the camera's audio reproduction. Audio is discussed in greater depth in Chapter 12.

Avoid wearing white or black because these colors increase contrast. Some people say earth tones are the best bet, but be careful not to blend into a brick background if you use it. Various shades of blue are also a good choice. Herringbone tweeds cause a moiré effect, which looks something like rainbows radiating from the clothes. It's best to stay away from tight patterns in general.

Jangling bracelets and earrings will be annoying on the audio track. If the character must wear them, you could tape them together. Otherwise, just be careful about rattling them.

Your make-up depends on your character, but a light base never hurts. Be sure to bring powder to your set to combat shine unless you want the sweaty look.

Keep your hair off your face, and remember that if you can't see the camera, the camera can't see you. In movies, acting happens in the eyes. If you find yourself blocked by either your own hair or the other actor, shift until you can be seen.

To decide whether or not to look into the lens, imagine yourself in front of an audience. If you would be talking directly to the audience, then it's okay to look into the camera. Otherwise, never acknowledge the camera. If you have trouble looking into the lens when your view must pass from left to right, try the "Burt Reynolds blink." (We don't know if he really does this or not, but that's what people call it.) At the exact moment when your eye would contact the lens, blink. Ever wanted to avoid someone's stare? Rather than make eye contact, you blink. The best way to deal with the lens, however, is just to lose your self-consciousness about the camera's presence. It is your friend. It is going to make you rich one day—and famous.

Try to get the scene all in one take. Decide on where zooms and close-ups should be when you do your preproduction, rehearsals, and camera rehearsal. Close-ups should include head and shoulders with one or two optional tight shots, losing the hair (metaphorically speaking) and keeping the chin. All camera work should be kept well within the camera operator's abilities.

Shoot as many takes as you need until you get one you like, but keep in mind the law of diminishing returns. If you're up to take 8 or 9, the performance may not get any better. In a situation like this, call a break, have some laughs, find some new details that will interest you in your character again, and then shoot some more.

If you are sharing with a partner, shoot the scene two ways. Shoot it once favoring you and once favoring your partner. An over-the-shoulder shot where you are shown full face is preferable to two profiles, so flip a coin to see who gets to star first, then reblock the scene for the other actor.

Unless you have booked a multicamera studio, shoot single-camera film style, and if you choose to edit, plan your cuts beforehand. For a simple and economical scene with good results, try four edits. The first should be an edit near the beginning to take you from a master shot to a loose close-up or a two-shot. The next edit occurs two-thirds of the way through to get your big, emotional speech in a choker, losing the hair and keeping the chin. (Watch how they frame these shots in the soaps.) A third edit gets you back to one of the above shots, and the fourth edit allows you to end the scene in one of the following ways: a master shot, a tight close-up for your final touching moment, or a loose two-shot for the big kiss.

If you're doing the shoot yourself or with friends, shoot on VHS, S-VHS, or Hi-8, and if you plan to edit, make sure that you choose an editing studio in advance that can edit in your format. The copies that you send to the agencies will be made on VHS.

Just as in the editing studio, it's important to keep the room cool, especially when using lights, so designate one person to turn the air conditioning off during a take and on after you cut. It's also important to post a sign outside the door asking for quiet and to unplug any phones while you're rolling.

During your preparation, take a moment to remember that you went into acting because it was fun.

And that's it! Turn off the air conditioning, and go for it.

14 Camera Operation

The purpose of this chapter is to help those who will use a camera to shoot a demo scene achieve the most acceptable, even professional-looking, results. Manufacturers continue to simplify controls and improve picture quality while reducing the size and price of their units. However, there is still plenty to be learned when it comes to camera controls and the smaller size can even work against a steady camera shot.

We have divided this chapter into two sections: Camera Controls, and Operating the Camera. These two sections will provide you with all the information you'll need to shoot a successful demo scene.

DEMO TIP: Read and understand the manual for your specific camera.

Every company that ever designed a camera has a slightly different idea of where the buttons should be placed and how they should work. If you rent a camera, be sure the manual comes with it, and before you leave the rental house, have the person you're renting from go over all the functions and operations with you. If other customers are lined up, wait around until the rental person is free. Don't arrive for your equipment right at closing time.

Generally, all cameras work in much the same way. They can run on alternating current (AC), which comes from the wall plug, or on batteries. Your preference should be AC because the battery will always run out right at your best moment. Count on it. If you are shooting outdoors or need the camera to run on batteries for other reasons, be certain that all the batteries (rental places call them "batts"), tested full before you left the rental house. If you're charging your own camera batteries, be sure you have done this properly once before. Follow the instructions in your manual, and carefully mark the batteries that are charged. Some batteries have little tabs that you slide to red for charged and black for discharged. Batteries may develop a "memory" if you discharge them partially to the same point over and over again. Instead of running for forty minutes, they will begin to quit after twenty. To avoid this, discharge them fully as often as you can, and try to avoid shutting them off and recharging them at the same spot over and over again.

When you rent a camera, always get both AC and batteries, and don't forget a charger. Take as many batteries as they'll give you without an add-on fee. You'll usually be provided with all the extension cords and plug strips you'll need. If you have conducted your location scout, you will know what your needs will be.

Camcorders and cameras with remote recorders are exactly the same as far as operation is concerned. The camcorder simply has the recording deck built in (on-board). Consumer cameras are sometimes equipped with all sorts of extras like a character generator, background colors, and certain effects that are not found on professional cameras, but when you get past these, you need to deal with only a few camera controls.

Camera Switches and Settings

Power Switch

DEMO TIP: If it doesn't work, it's not plugged in or it's not turned on.

When the camera is switched on, a little light will be illuminated somewhere and you will see light in the view finder. There may also be a standby switch, which allows the camera to be ready to shoot while conserving power. If the power switch is on, but the view finder is off, turn the standby switch from "standby" to "operate."

White Balance Button

Point the camera at something white in the direction you intend to shoot under the lighting conditions you are going to use, and press and hold the white balance button until the light in the view finder or other indicator stops flashing. For information on the theory of white balance, refer to Chapter 12.

Auto-White Switch

If there is too little light for the camera to white balance, set this to the picture of a light bulb for indoors or to the picture of a sun for outdoors. The camera may have a setting for completely automatic white balance as well, and you may use this if you are uncertain about white balancing yourself. Most cameras will "remember" a white balance setting for a certain period of time, usually from five minutes to twenty-four hours. Consult your manual for this information.

Filter Wheel

If you are using a professional camera, you will find a filter wheel at the front of the camera behind the lens. Set this wheel to the correct filter—32K, 45K, 65K, etc.—or to "closed." The closed setting works like an internal lens cap and should be used when the camera is not in use, particularly if it is a tube camera.

The Lens Cap

If a tube camera is pointed at a light source for too long, the tube will burn and will sometimes permanently retain a mark called a *burn-in*. If a burn-in occurs on a rental cameral while you are using it, you will be expected to pay to have the tubes replaced, which can cost more than a thousand dollars.

To cure a temporary burn-in, point the camera at a white card at the correct iris setting for twenty minutes or so. When using a tube camera, the best plan is to appoint someone to keep an eye on the lens and to always cap it if it is not in use. Although chip cameras can be pointed at a light source, most people avoid doing so unless shooting the light source specifically.

Record Button

Quite often you will find two record buttons on a camera: one behind the zoom control, to be operated by the thumb, and another on the front of the camera. Both will stop and start the recording.

When you are recording, a red light inside the view finder indicates that the camera is turning. Cameras usually have a little window on the door where the tape goes in. This allows you to look inside to see whether the tape is going around. Push record, then look in that window. If the little wheels are revolving and the red light in the view finder is on, relax—the shot is going down. Never forget about this light. Even experienced camera operators can occasionally get so caught up in the action (particularly if it's shocking or highly dramatic) that they stand there shooting away only to discover when everything is over that they never pressed record.

DEMO TIP: The camera will not record unless you push the record button.

Some examples from real life: A casting agent we know brought us a tape of a day's casting session to be dubbed and sent overnight to a large advertising agency in New York. The agent had seen and interviewed a hundred people. When we ran the tape, we found only three interviews on it. The rest of the tape was blank. After the third interview of the day, no one remembered to press record.

Want to hear another? Two guys started up a video company to tape events. Their first gig was a huge reunion at a Beverly Hills hotel, which people traveled from all over the world to attend. Those two interviewed everybody; they were very conscientious. They held the mike up to every guest. Yes, they had remembered to press record. Unfortunately, they hadn't remembered to plug in the sound.

Then there was the poor fellow who got things mixed up. He thought that the red light meant he wasn't recording. After taping the Fourth of July Regatta, he viewed the tape to discover that what he had was the world's longest nautical footwear commercial.

DEMO TIP: The monitor goes downstream of the recorder.

One more story: Some producers decided to do a multicamera taping of a rock event at a large downtown theater. Being very frugal, they hired an outfit we'll call Squeaking Wheel Video Production, who had turned in the lowest bid. The bid was so low, in fact, that the producers began to suspect that all the other companies who bid had been out to cheat them. Once the concert was over, the producers were so excited about how well it had gone that they raced over to our place the next day, with three hours' worth of tape. When we put the tapes in the machine, we could hear the music, but we couldn't see anything. We tried the tapes in a different machine, but there was still no picture. It took quite a while for the situation to sink in but the unalterable fact was that no video had been recorded on the tape.

This is how it happened. SWV Productions had saved money by switching live and recording the concert with a single recorder. They did not use isolated decks, and they did not run a second recorder as a backup. They had plugged their monitors in upstream of the recorder. If they had plugged in at least one monitor downstream of the record deck, they would have realized that no one had run video cable from the switcher to the recorder and no picture was going onto the tape. We couldn't even use the audio. Without picture, there's no sync. Without sync, the sound wobbles all over the place.

It's very important to press record. It is equally important to record a test and play it back to be sure you're getting what you think you're getting.

Eject Button

When you push the eject button, the little door where you insert the tape opens up. This little door is referred to as the *cassette elevator*. If you've done absolutely everything right and you still can't get the camera to record, press the eject button, remove the cassette, and check to make sure the write-protect button or tab is in the correct position. Once it is, it will.

DEMO TIP: The red button tab or gate must be in the correct position for recording to occur.

As discussed in detail in Chapter 18, to record on three-quarter-inch, the red button must be in. For VHS, the tab must be in. For Hi-8, the red door must be *open*. If the VHS tab has already been pulled and you want to continue recording, put transparent tape over the opening. If the above is not done, the camera will not record.

Zoom Control

The zoom is a rocker switch located on the right side of the lens. Some are single speed; others are variable. Beginning camera people immediately fall in love with this device. They hate focusing, but they love zooming.

DEMO TIP: Be sure to motivate a zoom.

The motivated zoom can become second nature. Think about the following:

- If the subject becomes animated, zoom back.
- If the subject becomes introspective, push in.
- If another person enters the frame, pull out.
- To pick up a detail—for example, a butterfly landing on your subject's hand—zoom in.
- If a large object, like a steamship, enters the frame, pull back.
- To begin a shot, start tight on the flower, then widen out to reveal subject.
- To end a shot, zoom in on the subject's tear.
- To begin a shot, start wide on the whole skyscraper, and zoom in as the subject comes through the revolving door.
- To end a shot, pull back from the couple to reveal the sunset.

DEMO TIP: Before you record, zoom in to focus, then pull back to frame your shot.

Until you have a very practiced eye, always zoom in to focus. You can do this manually to speed things up and save the batteries. A fuzzy, slightly out-of-focus shot will drive the viewer crazy. The eye tries to compensate, and that just can't be done. Practice pulling focus in your spare time without the camera rolling.

The Focus Ring and Focusing

Before working with focus on a video camera, slide the built-in lens adjustment lever to the sharpest and most comfortable position for your eye. Once you are satisfied with its placement, you are ready to being focusing your shot.

The sharper the edges of the object on which you must focus become, the closer you'll be to optimum detail. In the view finder, the edges may even appear to have a thin outline once they are exactly right. If your subject is a person, focus on the pupil of his or her eye. If you're in so close that only one eye can be in focus, the one to choose is usually the nearer one. Bring the edges into focus until they are crisp, then leave it. The more light you have on the scene, the easier focusing will be and conversely, the lower the light, the fuzzier your picture will become. As in all photography, the less light you have, the narrower your depth of field will be. Depth of field is the part of your scene between you and the horizon on which you are able to maintain a sharp focus.

There is nothing that will give the amateur away more quickly than turning past the point of perfect focus, and then coming back. Some beginners do this many times in succession, some never do get it right. The reason people don't like to focus is that it takes skill and practice and it's sometimes hard to be sure when it's right. A certain amount of self-confidence is needed to get focus correctly. Stick with your decision.

It's important to think about focus when discussing camera controls because the view finder you choose greatly affects the sharpness you are able to achieve. A two-thirds-inch consumer quality view finder will be far less precise than a one-inch professional type, and your ability to achieve a crisp and detailed focus will be affected no matter how practiced an eye you have.

Using a monitor to focus, particularly if you are using the consumer type of view finder, is recommended, and if you have one connected, you will benefit from using it. The higher the resolution of the monitor, the more control you will have. A black-and-white image will improve your ability to attain a precise focus whether you are looking at a monitor or into a view finder.

One day you will start to pull focus as you zoom. You will be able to see detail in the view finder, and as it starts to get soft your hand will compensate by turning the focus ring. Clockwise for nearer objects, counterclockwise for objects that are farther away. That's clockwise-nearer; counter-farther. One day it will be automatic. It'll be such a thrill the first time you realize you've learned to zoom and "rack focus" simultaneously that you'll want to tell strangers.

Focus pulling is an art that requires exceptional concentration and precision. For every motion picture camera, there is a professional focus puller who has dedicated his or her life to defocusing on the guy and sharpening up on the air, just at the moment the guy says "Let's get married, whaddya say?" The men and women who do this job are unsung heroes. They are also expected to schlep the camera.

Auto-Focus Switch

The focus switch can be set for automatic or manual operation. Most professionals don't use auto-focus. Until recently, the subject had to be framed dead center, making the rule of thirds impossible. The rule of thirds is an artistic convention that says you should mentally divide the view finder into thirds horizontally and vertically. Where the imaginary lines meet, like the corners of a little box, is where the subject should be framed, rather than dead center. This adds a sense of movement and perspective to the shot. Try it for yourself, and you'll see how it works. Don't employ this to the exclusion of all other frame composition; just keep it loosely in mind.

Another auto-focus problem arises when a secondary actor crosses the frame and the camera refocuses on that person, while the viewer's attention should remain on the main subject. A last annoyance is that the camera continually gets fuzzy as the auto-focus searches the field to correct itself.

Manufacturers have been working hard on these problems, and the auto-focus feature is getting better all the time. If you go out to buy a new camera, pay particular attention to how the auto-focus works.

Auto-Iris Control

All video cameras have auto-iris, similar to an aperture on a still camera. Most people, including professionals, use it regularly. If you are shooting to edit, however, it's a good idea to set the iris manually to make sure that the light levels you record remain consistent. Another reason for manual iris control is to get enough light on a subject standing in a strong back light, such as a person against a sky. In his case, you can iris up, opening it until the face is clear, allowing the background to be a little overexposed. If you use the manual iris, be careful not to allow "hot," or bright, areas of the picture to bloom, washing them out entirely.

Some consumer cameras have a three-way switch on the auto-iris. It can be switched from back light to normal to highlight. Highlight is used when taping a dancer in a spotlight, for example.

Recording Speed Switch

Some cameras and recorders offer a choice of taping speeds. Always tape on standard speed if you plan to copy or edit the tape. As mentioned earlier, slow speed drastically reduces picture quality, and professional editing machines will not work with this speed.

Review Button

Most cameras have a review button that, when pressed, rolls the tape you have just shot back a few seconds to allow you to see what you've got and to ensure that it's there. For more on this topic, see the discussion of electronics to electronics (E to E) in Chapter 18. When you look through the view finder during recording, you do not see what's on the tape. All you see at this point is your shot transferred electronically to the view finder: electronics to electronics. To see what you recorded, you must watch the tape back. Some professional cameras use an extra set of heads that read what is being recorded on the tape and play it back through the view finder. They are known as *confidence heads* because you see what is recorded as it records. If there's a bad piece of tape, you will spot it.

Using the review function, after replaying the last few seconds, the tape will automatically park so that recording will begin exactly where you stopped the tape, making a clean camera edit. The tape parks at a preroll point, usually two seconds, so be sure you are actually in record before you yell, "Action!" The best procedure is to record ten seconds of tape before the action begins to make sure the editing machines always have enough preroll time when you're cutting the tape together.

Shutter Speed Function

Newer cameras have a shutter speed function. If your subject is moving fast, as in sports, the picture can blur at normal shutter speed. To compensate for this, you can raise the shutter speed. When you do, you will also need to open the iris because increasing the shutter speed reduces the amount of light reaching the camera. When making a product demo of a fast-moving object or machine, experiment with the shutter speed to obtain the best possible picture. Be sure you have plenty of light.

Operating the Camera

Now that you know how it works, let's talk a little more about how to work it.

Tripods

DEMO TIP: It doesn't matter how you do it, as long as you keep the camera still.

To ensure that the camera will remain motionless during shooting, use a fluid-head tripod, and make sure that the camera is tightly anchored to the camera plate. If you don't have a fluid-head tripod, use whatever you have. If you don't have anything, set the camera on a table. Practice making smooth tilts and pans. Stand away from the camera so it will not move in time with your breathing.

If you have focus changes to make, once the actors have their marks, stick a little piece of masking tape on the focus ring, and mark your appropriate focus points with a pen. You can even number them so you know which one you're going to next. The monitor or the view finder is the final arbiter, however. If the actor misses his marks, don't leave him out of focus just for spite, okay?

DEMO TIP: Less is more.

Imagine that your job is to record the scene, not to run the camera. That purpose is often best served by locking the camera down, making sure it's rolling, and standing to one side. Plan zooms ahead of time, and stick to the plan. If you find yourself having to tilt and pan a lot, try a wider shot, and keep the camera still. The farther out on the lens you are—meaning you have zoomed to telephoto—the more likely the camera will shake, and the more obvious the shaking will be. Instead, reposition the camera closer to the actors or to the subject.

DEMO TIP: Beware of fancy camera angles.

Keep the camera angles simple. The camera should be placed at about the actor's eye level—a little higher if you want to give a feeling of domination over the actor, a little lower if you want the actor to seem more powerful. If the actor is seated for most of the shot, bring the camera down to eye level. If the actor is standing, raise the tripod. After the shot is done and you know it's in the can, if you can't resist going for something arty, try putting the camera on the floor.

DEMO TIP: Watch your frame like a hawk.

Continually and consciously watch the edges of the frame. Is there a light stand creeping in? Are you off the set? Is one of the actors slowly disappearing? Is there too much head room? Is the top of an actor's head getting cut off? Can you see all the props you're supposed to see? Are the actors centered in the frame? If the actor is looking to the left, is there lead room on the left? That is, have you panned a little so the actor is on the right side of the screen when their look is left, or vice versa? This convention prevents the viewer from feeling cramped up against the edge of the picture.

The Hand-Held Camera

DEMO TIP: Brace yourself.

If you need to do some hand-held camera work, use whatever you can to keep the camera steady. Lean against a fence or a pole. Hold the camera with two hands: one on the zoom and the other bracing the camera from underneath, leaving your thumb and forefinger free to pull focus. Keep your elbows tight to your sides—not rigid, just braced. Stand with your feet shoulder width apart, stiffen your toes, and dig them in as if they were claws. You are the tripod. Now allow the camera to "float" securely on your shoulder. Let the movements be fluid.

If the edges of the frame begin to shake, widen the shot. If your arms start shaking because they're tired, turn the camera off, set it down, and hold your arms up in the air for a while. They'll settle down, and in several seconds you can go back to work. (If you're handling a product in a shot, by the way, and your arms tense up, the same procedure will work to keep your hands from shaking.)

Let the camera rest on your shoulder; don't try to bring your shoulder up to the camera. If you need to walk with the camera, bend at the knees and flow along, keeping the shot wide. When doing a lengthy pan, point your feet toward the spot where you plan to end the pan, then twist at the waist back to the beginning of the shot. Slowly unwind yourself until you finish where your feet are pointed. Learn to open your other eye so you can see where you are going. This way you'll know when something is about to enter frame. This is one of the secrets of smooth camera work.

DEMO TIP: A camera that fell in the water with you is probably not a camera anymore.

Remember that your right side is blind. If you are close to some physical danger, such as traffic or a cliff, have an assistant hang onto your belt and watch out for you.

Buying a Camera

Advice on buying a camera could fill an entire book. We suggest following the same basic procedure as when buying a VCR (refer to Chapter 3.) VHS cameras abound on the market and they accomplish the compact version by using the VHS-C twenty-minute tape load, which will then play in full-size machines with an adaptor. If you're planning to fall in the water, Sony puts out a nifty little yellow 8mm that's good to several fathoms. Still photographers might consider a Canon L1 Hi8 capable of handling a variety of lenses. The Sony EVO-9100 is still going strong, and its time code feature makes it an ideal professional single-chip camera. If you plan to use a Steadicam JR, described at the end of this chapter, the camera must be four pounds or under. The Sony 9100, stripped of all its extra gear such as the view finder (the JR has a built-in monitor), and using the flat NP-55 battery, just makes it.

When you outgrow the cameras available in consumer outlets and head for television studio equipment dealers, you'll find the two- and three-chip cameras produced by JVC, Sony, Panasonic, and others to all be of exceptional quality. Your choice will be governed by matching your format to your other equipment and your budget.

Handheld Supports and Stabilizers

Various and sundry hand-held supports are flooding the market, all based on a different idea, and it is impossible to rate them. Try out various supports in the camera store, and if one really works for you, go for it. There is one that has gained very special attention, however.

The Steadicam JR. About fifteen years ago, Garrett Brown invented a device that stabilized hand-held cameras and won an Oscar for its design. It is large and heavy, and it costs from $20,000 to $40,000. In 1990 he brought out the Steadicam JR (pronounced "jay are"), which is based on the same principle, yet is so simple in design that it can sell for $600. Made by Cinema Products of Hollywood for cameras that weight under four pounds, its operation can be mastered in a couple of days. If you are considering a moving shot of any kind, the Steadicam JR will make your hand-held shot look like you used a dolly or a crane. This device can be rented for a few dollars a day, but it will definitely require practice before you use it. If you do real estate videos, this device should be an essential part of your gear. Once you master it, your work will look wonderful!

Part III

Technical Considerations

15 Getting the Best Picture Quality

Broadcast-Quality Video

There is no more overworked phrase in video than *broadcast-quality*. Almost every video equipment advertisement in the Yellow Pages contains it. Someone said that if *broadcast-quality* has to be stated, then it isn't. This is what it is. The Federal Communications Commission (FCC) created a set of technical standards that any recording or transmission must meet before it can be broadcast over the air. Cable, which didn't exist when the rules were made, is exempt. These standards are measurable on the waveform monitor—an oscilloscope that measures various elements of the video signal, such as brightness and black level—and the vectorscope—which is used to monitor color. Without these two instruments, it's impossible to tell whether a tape is broadcast-quality or not.

The waveform monitor (WFM) measures many things, but one of the most important for your purposes is the amplitude of the brightness—known as *luminance*—which must peak at 100 IRE (a unit named after the Association of International Recording Engineers). If the luminance goes above this level, the picture is said to *bloom*, and you will see a patch of white in the picture where all contrast and color are gone. Bloom can occur on someone's cheek, on a white shirt, or in a bright sky.

The WFM also measures black level, which, to be broadcast-quality, must reach its lowest point at 7.5 IRE. If it dips below this level, the picture will be very dark, and portions of it could be blacked out completely. It's a good idea to ask your editor if the source monitor is upstream or downstream of the time base corrector. If it is upstream and it has been matched to the program monitor, it will show you the picture before it is corrected by the TBC. This allows you to compare that picture with the program monitor. If the latter is darker or brighter than the source and the editor hasn't noticed, bring the discrepancy to their attention. Keep in mind, though, that editors have many things to do, and they can't do everything at once. If they're working on the sound, wait until they're working on the picture to mention it. If the source monitor is downstream of the TBC, you must rely on the WFM alone, so politely ask if luminance and setup—another name for *black level*—are okay.

The vectorscope measures hue, telling you if the colors are too green or too red or just right, and flesh tones will be the most obvious. Saturation tells you if the colors will bleed into each other, be washed out, or turn out just right. Hue is set by putting the color bars on the screen for adjustment. After than, hue can be adjusted by eye to get the flesh tones right for each different source tape. Simply match the flesh tones to the preceding scene, and every so often scan through the tape to make sure the hue hasn't shifted. If you're unsure, go back and watch through the tape, checking hue (and nothing else). If it's a long project, such as a collected works tape, go back to the color bars every so often.

Saturation can be adjusted as you go by making sure the circle of light on the vectorscope screen does not spread out beyond the little boxes or fail to reach them at all. The vectorscope has a pattern on its screen that looks like a constellation in a star chart when color bars are displayed. There are seven little hollow squares grouped in a circle, one for every primary color. When the color bars at the head of the tape are played, the editor adjusts the TBC until the bright little dots representing the colors are all in their respective boxes. Once this is done, the colors will be broadcast-quality.

If you have rented a studio to do a two- or three-camera shoot, the cameras must be matched by the same methods. This is done with a device known as a camera control unit (CCU), which performs much the same task as the TBC. If the cameras are not set up or matched prior to the shoot and monitored throughout the day, you will end up with a color shift. That is, the blue shirt you have on in one angle will be green in another. This can be fixed in postproduction with the TBC, but it's a nuisance. To check that the cameras are matched, have the technician who is doing the switching dissolve from one camera to the other with the cameras set to color bars. If the little pinpoints of light are in almost exactly the same spots in their little boxes on the vectorscope, your cameras are matched. The respective displays on the WFM must be matched as well.

You can elect to do a two- or three-camera ISO shoot, using isolated cameras—that is, each camera records on its own separate recorder. In this case, final corrections have to be made in postproduction. Even with a WFM and a vectorscope on the shoot, however, the color matching and broadcast quality could be questionable.

Some people will tell you that three-tube or three-chip cameras are required for broadcast-quality. This was once true but no longer. As you can now see for yourself during the news, consumer footage ends up on the air almost every night.

Sony's single-chip camera, the Sony EVO-9100, has been nicknamed "the disposable camera" by news crews. We're told it was this camera that got the incredible footage in T'ien-an-Men Square. The American newspeople achieved this by shooting the situation, pocketing the little 8-millimeter tape, tossing the camera, and getting the blankety-blank out of there.

Today, these popular little cameras are found in every news truck, and if an event is too dangerous or too immediate to get with a regular news camera, they shoot it with a 9100. Though the resulting picture isn't quite as good as the full-size 3-chip cameras, we seldom notice when we're just watching television. The camera sells for a little under $3000. They are usually not sold in consumer video stores, but dealers can be found in the Yellow Pages under "Television Studio Equipment."

Most companies market "industrial" cameras with one, two, or three charged coupled device (CCD) chips. Many of these cameras are true broadcast-quality. If you're in doubt, call a dealer or the manufacturer and ask.

There are many other requirements for broadcast-quality video, but these are the most important ones for demo tapes. If your tape is not going to be broadcast, think of broadcast quality as a guide, but don't be overly impressed by the phrase. A studio that does not guarantee full broadcast-quality will be far cheaper and just fine for your purposes if you're on a budget. If you are not using full broadcast-quality, bear in mind that what you see is what you get.

If you are really to get what you see, when no vectorscope is present the monitor must be adjusted correctly. If the studio can generate color bars, put them up on the screen and adjust the monitor until the red is red, the magenta is magenta, the blue is blue, etc. If a gray scale is available on the bars, it will be in the lower part of the screen. Adjust the monitor until the gradation is apparent. A monitor that is out of adjustment can allow the operator to get the hue completely wrong throughout the tape if the colors are adjusted by eye. When you play your copy at home, everyone on it could be green.

Common Technical Problems

Monitor adjustment is just the first danger of trusting what you see without question. Many other serious problems sometimes cannot be seen until a copy of the tape is played in a different machine. But don't despair; the next two demo tips will help eliminate 90 percent of these potential disasters. When confronted with a machine or a tape problem, the first culprits to consider are overheating and clogged heads.

Overheating

DEMO TIP: Bring a sweater.

If you've ever been in an edit bay, you've probably noticed that it was chilly. An edit bay or studio must be kept cool for the machines to function correctly.

If a machine gets too warm, the first thing that will be affected is tracking. This may not be noticed until you go down a generation. Then you will discover that the picture jumps, jitters, or seems to move up and down on the screen. You may discover faint crisscrossing lines at the top or bottom of the picture on all the scenes that were done when the room was too hot. "Uh, oh," will be about all you can say at this point. You are going to have to go back and recut. What has happened is that heat has loosened or stretched the tape, and it has wobbled around during recording. Keep in mind that the tape is flowing around a drum that is spinning at an extremely high rate. It is an unstable medium being asked to perform an outlandishly precise task.

Other nasty occurrences can often be traced directly to heat. If the editor suddenly has trouble getting accurate edits, check the room temperature. If the picture continuously jumps at every edit, same thing. The last heat-related problem and worst of all, happens when the tape leaves the heads altogether and gets jammed in the pinch roller. The machine at this point is said to have "eaten" the tape. This is the true meaning of the word *glitch* in an editing studio. When you play the tape back, you will see lines of snow on the screen, hear a terrible grinding noise coming from the VCR, and your editor will immediately burst into tears. You have two choices. You can start over from the beginning, or you can copy your work onto a new tape, losing a generation, because the edited master you have in that machine is now history.

When the machine suddenly stops, a red light begins flashing, and the editor, who may not yet be in tears, is suddenly as alert as a cornered antelope. This also indicates that the tape has jumped the tracks, so to speak, and the tape transport may have taken a bite out of it. After shutting down the machine, taking a break, and cooling the room, if you play the tape back and it looks fine, make a copy of that section on a scratch tape, and play it back on another machine. If everything is still fine and there are no perceptible jumps, you can go back to work and consider yourself one lucky demo producer.

The reason for the sweater is simple. Editors are very sensitive and sympathetic people. If you sit next to them shivering, even if you don't say anything, they will get up and adjust the air conditioning, promising themselves that they'll remember to turn it up again in a few minutes. Then both of you get very busy and they don't think about it again until . . .

Clogged Heads

DEMO TIP: Make sure the heads are clean.

Oxide on the heads and a dirty tape transport—the path that carries the tape on its journey through the machine—are the second most common cause of tape problems. When an editor is on a deadline or the previous appointment has run overtime, it's a great temptation to skip cleaning the machines. As a client, you will do yourself and your blood pressure a great favor by understanding that, like your dentist, doctor, and motor vehicle examiner, editors get behind. Often it's not their fault; clients regularly underestimate the amount of time they will need. If it's a matter of running overtime by fifteen or twenty minutes, editors find it difficult and unpleasant to ask someone to leave and come back another day.

When you finally enter the edit bay, simply let the editor know that you can wait a few more minutes if he or she needs to clean the heads before getting started. This will impress editors with your understanding of their problems, allow them to relax (because they will not be scolded for making the client late), and you and your editor will be on the same wavelength right from the start. The whole session will go faster. You will save money, and the editor will do his or her best on your project.

Even if your editor declines because the heads were just cleaned or because this particular house cleans them at regular intervals, you will still benefit by having set up a sympathetic relationship. You will also have satisfied yourself that you are probably working on a clean system and that you will not be held up by oxide-related problems.

What are oxide-related problems? The first indication of a problem is that the picture starts to become washed out, then it becomes snowy, and finally it "goes to snow" altogether. When this happens, your inclination may be to point at the monitor and scream, "There's nothing on my tape!" Don't overreact too soon. You'll need to hang on to this tip even tighter now that you're sitting in an edit bay. Your tape is almost always fine. It has not been erased. The heads of the machine have become clogged and cannot read it. All will be well once the heads have been squirted with a touch of Freon and carefully cleaned with a Technowipe—a little square of lint free cloth that costs about as much per foot as a yacht.

Other symptoms of dirty heads or a dirty tape transport can be sync roll (when the picture rolls on the screen as if the vertical hold on your television set needed adjustment). You may also experience uneven tracking, or inaccurate edits.

Cigarette smoke seems to be attracted to video heads. If you or your editor smoke and smoking is allowed in the room, clean the heads often.

The usual reason for the clogged heads—assuming you began with clean heads—is an old source tape. The binder (glue) has begun to lose its grip on the oxide, which now comes off on everything. Actors run into this problem more than most people because they need to use source material that has been around for a number of years or because the tape that came from the production office is an old or overly worn tape. Producers of product and service demos could encounter this problem if they buy used tape. You can run into so many problems with used tape that you'll spend far more than your savings in the time wasted fixing problems.

This problem can often be solved by transferring the old source tape to a new tape by making a straight-across transfer without a TBC or anything else in between. If the signal quality has deteriorated too far, this won't help, but it's always worth a try. Once your source tapes are approaching ten years old, make an appointment to have them transferred to new stock while the picture and signal are still reproducible. Today's tapes have a much longer life than older tapes, and it's said that they will last indefinitely. The dub you make will become the new source tape, but don't throw the original away.

The quickest way to age a tape is to leave it in your car or trunk on a hot day. Before we put a tape in a machine, we touch it to see if it's hot. If it is, we cool it down first. Some tapes can get so badly heated that the cassette warps and is difficult, sometimes even impossible, to insert. Tapes are like kids: Don't leave them in the car when you go shopping.

An editor may refuse to work with certain old tapes that may already be stretched and worn because there is a strong possibility that the machine will eat the tape. This is very bad for the machine, and if the source material on the destroyed tape was irreplaceable, there is always the danger that the client will try to blame the editor when it isn't the editor's fault at all. If your tape is refused, ask about the possibility of a straight-across transfer; the editor will usually agree to that.

Bear in mind that these solutions for overheating and clogged heads are just the first ones to try if you encounter a machine or a tape problem. Before you and your editor decide to shut the system down, cancel the session, blow the deadline, and await the team of electronic surgeons from the factory, cool the room down and make sure the heads are clean.

Handling Technical Problems When They Occur

It can't hurt to repeat a favorite tip that applies to the edit bay: It's not an airplane; it can't crash. This means that you have time to look first for the most easily remedied diagnosis of a problem and then work your way up in difficulty. Very few situations do anything worse for the client than waste time or anything worse for the editor than waste money. So let your editor take the necessary time when thinking through an abnormality.

Machines do fail, however, and if one does, it can exhibit any of the symptoms previously discussed and may need engineering attention. Try the simple solutions first, then look deeper. For instance, many bewildering events are corrected by simply turning off the machine, waiting five seconds, then turning it back on. Who knows why? Who dares ask? Sometimes when the edit controller seems to go berserk, it's because it got a whack of static electricity from someone's hand. The solution to this problem is to mix one part fabric softener with one part water and to spray it on the carpet around the machines. Finally, there are a lot of fuses in an edit bay, and they are instantly replaceable. Usually, however, if it doesn't work, it's not plugged in or it's not turned on.

The wiring system behind the edit console is lovingly referred to as "spaghetti," and wires often need to be switched for a variety of reasons. If you can't get a picture or the picture you want, your editor may suddenly disappear out of sight behind the console. He is looking for the cable he needs to replug. This is part of the whole editing process. If it wasn't necessary occasionally to "swap a cable," you would not have the wide variety of effects that are available to you. In fact, a knowledgeable editor can coax or trick the machines into doing much more than the manufacturer intended, often by swapping a few cables.

The edit console has dozens of switches and buttons. If one of these is in the wrong position, the screen may go blank or the effect may not occur. The editor will have to trace through the system to find the button that needs to be switched. If only one of these glowing little cough drops is out of place, everything can stop. This happens now and then in a session, and it's normal.

If unexplained holdups persist, however, that's a signal you may have been assigned a trainee. One sure indication is when the person slowly becomes motionless, stares fixedly at the monitor with a hand poised over a keyboard, and appears to be sweating bullets. If you're an actor and you've ever totally dried on stage, you know what the editor is feeling. Unfortunately, now is the time to ask some questions. Another sign is when the editor constantly runs out of the edit bay, returning only to stare at the system once more. Ten to one, the editor is consulting someone out there and learning on the fly.

DEMO TIP: Don't burn your bridges.

Your objective is to get your project completed. It's far simpler to complete it where you're seated than to start all over at another studio. Stay calm, and stay on the editor's side. The following comments could help. "This machine doesn't seem to be too user-friendly, does it?" "Are we having a problem that someone else here might be more familiar with?" "What does the manual suggest?" "Do you want me to take a little break and leave you alone to sort it out? Just let me know when we can get back to work, I'll be out here." If the situation remains stalled, let the boss know that progress seems to have ground to a halt and ask whether a more experienced operator can take over.

DEMO TIP: Allow your editor to get a second opinion.

It sometimes happens that even the most experienced and professional editors develop a mental block. With the intense concentration required for this job, a mind can just go blank. The best thing for the editor to do in such a case is to invite another editor to come in for a moment and look things over with a fresh eye. Chances are good that the problem will be solved quickly, and you'll be off and running. We used this method all the time, and it saved lots of time.

However, if nothing gets you going again and you're certain that nothing will help, explain the situation to the boss and ask for an adjustment on the bill. Gather up your tapes, and tell them you'll return when they've got the problem sorted out. In the unlikely event that a misunderstanding becomes unavoidable, stay cool and repeat your requirements civilly. The most important thing is to get your tapes back under your control.

DEMO TIP: Foresee the unforeseen.

When thinking about an upcoming editing session, human nature being what it is, it's easy to imagine everything going perfectly. In your imagination, you will not have forgotten any tapes; you will not have been held up by an accident on the freeway; you will not have thought about an unnoticed crease in your head shot that, after lighting, looks like your make-up has a crack running down the left side of your face. Nor did you know your main VHS source tape was recorded on slow speed and that copying it will put you forty minutes behind.

When you thought about your session, every edit went down the first time. Now that you're in the edit bay, you discover that you have to preview almost every cut to adjust the audio and the TBCs. Machines can and do break down, and if you're running *really* late, that's when it will happen. As stated by Murphy's Law, if something can go wrong, it will. In the edit bay, it goes a little more like this: If something just absolutely must not go wrong . . .

It's fine to imagine how quickly you can get through your session, but do your nerves a favor and also think about how long things could take. Then book the session far enough in advance to leave yourself an out. If there is a machine failure, you will need contingency plans. Remember, if you and your editor are not rushed, you'll have the leisure to think clearly, and your session will go like a knife through butter.

16 Defining Tapes and Untangling Formats

Defining Tapes

Before discussing the technical aspects of an actual editing session, let's define the tapes themselves.

The *source tape* is the tape that contains your original material. This is the tape from which you will edit. The *edited master* is the tape to which you will edit the selections of your material taken from your source tape. Once edited, the edited master is referred to as the *master tape*. This is the tape from which copies will be made.

A *copy* can be a copy of any sort of tape. It is unwise to make copies of copies because generation loss will occur each time until the picture finally becomes unwatchable. *Copy* is a synonym for *dub*, *transfer*, and *dupe*.

A *dub* can be a copy of anything, but it's usually a copy of the master tape. The process of making dubs is known as *dubbing* or *transferring*. Clients often phone a dubbing house and ask for a *transfer* of a master tape. This word can be a noun or a verb, depending on context. When referring to a small number of copies, the word *dub* is generally used. If you are making a large number of copies, the resulting copies are known as *dupes*, from the word *duplication*.

In a duplicating operation, record machines are said to be *slaved* to the source or playback machine. Thus, some people refer to the resulting copies as *slave tapes*.

A *submaster* is a copy of an earlier master that is used as an updated master.

A *scratch tape* is any tape the studio uses to record something when two or more passes are required to perform an effect. Scratch tapes can be used for laying down a music track or for shooting stills, for example. They are sometimes used as a B roll when a dissolve is employed. *Work tape* is another name for scratch tape. You are normally not charged for the use of a scratch tape unless you want to take it home afterward. If the studio where you're working does charge, ask whether you can provide your own. If that's not permissible, bite your lip and get on with the job.

In an A/B roll editing system, the *B roll* is a dub of a source tape known as the *A roll,* created so that a dissolve can be made between two scenes on the same tape. If you want to dissolve from one scene to another on the same source tape, you need to create a B roll. Check to see that generation loss is not too apparent. If it is, forget the dissolve and instead use a cut or a fade. These terms also refer to any tapes that are rolled in the A machine or the B machine. The B roll is sometimes a tape of another angle in a multicamera shoot.

Untangling Videotape Formats

Beta, often referred to as half-inch tape, has two distinct meanings depending on where you use it. The original word evolved like this. Once upon a time, a relatively small company introduced the first consumer video cassette recorder to the market and it used half-inch tape. The company was known as Sony and they called their machine a Betamax. Consumers soon abbreviated the name of this machine to "Beta" and it lasted long enough for everybody in the whole world to learn it. Before we go on to the second meaning of "Beta," let's discuss Beta's arch rival, VHS.

JVC, which stands for Japan Victor Company, introduced *VHS*, an acronym for Video Home System, often referred to as half-inch tape, which should not be confused with VCR, video cassette recorder, or VTR, video tape recorder. This format came out a year or so later than Sony's Betamax. They had used the time to iron out the early bugs from their system and even though Sony's machine had a somewhat superior picture, the VHS format was more user friendly. Video buffs went for "superior picture" but the average user went for "friendly." After a fifteen-year epic battle between the two format giants, Sony, who had long since become quite user friendly, finally threw in the towel by producing a VHS machine of its own and the tape to go with it.

Sony had a lot invested in half-inch tape and the Betamax format and it made sense to continue using what was already in existence. They therefore chose that format for their revolutionary broadcast system. They called this new system the *Betacam,* sometimes referred to as half-inch tape. Naturally, everyone who chose the new system began calling it "Beta."

This tape moves so fast that it whips through the equivalent of a two-hour length in 20 minutes. The cameras that perform this feat cost $15,000 apiece and more. If you have some shows you want to edit on the old kind of tape, Betamax, be sure to describe it very carefully to make sure the facility still owns one of those machines. If you're not sure which is which, just ask about the hourly rate. The difference between Betamax and Betacam will be immediately apparent.

MII format, sometimes referred to as half-inch tape, is Panasonic's and JVC's answer to Betacam, and it works on the same principle. In this case, the format shares its size with VHS.

Three-quarter-inch has been the workhorse of industrial video from the beginning of video. Also known as *U-matic* in deference to the way the tape is loaded around the heads, three-quarter-inch is the tape on which most professional actors' work is stored and, at present, most of their edited masters are on the same format. Think back to the news crews of the recent past, inseparably strung together. They were shooting on three-quarter-inch. One had the camera, and the other ran along behind with the giant "portable" recorder. New three-quarter-inch machines have been introduced by Sony called U-matic SP, and these offer a considerable improvement in picture quality as well as a reduction in generation loss.

JVC, however, has dropped the format in favor of *S-VHS*. Hi8 from Sony is now firmly established and these should serve us for the immediate future. Three-quarter-inch machines will be needed for a long time yet, though, because so much of our archival material is stored on the format. Some say three-quarter-inch has a maximum of ten years left, and that sounds pretty accurate to us, but if there's some old footage (other than home video) from the seventies and eighties that your kids decide to dust off, they'll find it on three-quarter-inch tape.

Two-inch tape created a legion of rugged individualists known as the Yardstick and White Glove Boys. They were editors who accomplished the edit by finding the picture they were after, then measuring eighteen inches down the tape to where the matching audio was and, blind as bats, cut it with a razor. They found the next scene the same way and spliced them all together. The white gloves kept the oxide off their hands and their fingers from smudging the oxide.

Two-inch tape is obsolete except for the odd soap that's still shot on it. All you can say about two-inch days is, "Boy, those machines were big!"

One-inch tape is still the workhorse in broadcasting. There are a few versions of it, the most common being one-inch type C. It is seldom used in the field and is being inched out everywhere except at the big studios by Betacam and MII.

D1, *D2*, and *D3* are up-and-coming digital broadcast tape formats that are having considerable impact on the industry. D2 has been in use since 1989, and D3 is now on the market. Because the video signal is recorded digitally, it is possible to go more than twenty generations without a significant drop in quality, and problems such as dropouts and clogged heads are automatically corrected. Special effects that once required two or more machines can now be done with a single machine, and the format has four digital audio tracks providing compact-disk-quality audio recording. This format is based on the three-quarter-inch U-matic system.

17 Edit Bays and Editing

Throughout the book we have organized our topics in the order that you will probably encounter each one, rather than in mechanically logical groupings such as everything about sound, everything about editing, everything about electronics, and so on.

We have approached this chapter in the same way, and so we discuss preparing a tape for editing in advance of the two types of editing from which to choose. We deal with the theory of editing at the moment you will need to apply it and possible technical problems at the point you will run into them. Tracking is an early problem, for example, while wrong-field edits will not occur until after you are into the edit session proper.

The last major topic in this chapter is a description of the machines and their function and these, while grouped together, are also described as closely as possible to the order in which you will encounter them. Let's begin with the first question you are likely to be asked by an operator, who will say, "Are we going to be working on-line or off-line today?"

On-Line and Off-Line

There are two types of edit bays: on-line and off-line. Much confusion exists concerning the difference between on-line editing and off-line editing, but the concept is absolutely simple. If the tape you are editing is going to be your finished product, you are working on-line. If you are doing a preliminary edit on a smaller, cheaper format and will cut your final tape on a higher-end format in a higher-end studio, you are editing off-line. If you have created your edit decision list (pro talk for "logged your tapes") at home or assembled your program on a cuts-only system before going on line, you have been off-lining. You then go into the on-line studio to create your final program.

The confusion of these ideas really arose from the fact that high-end studios with high-end prices are normally where on-lining is done, and many people assumed the phrase was related to the sophistication or broadcast-quality of the product. In actuality, it is simply whether or not the tape you are working on at the moment will be the finished product. In demo tape editing, you are practically always on-line if you are in a studio.

Preblacking and Time Coding

Preblacking—also known as *edit black*, *base black*, *crystal black*, *prestriped*, and *coded*—refers to prerecording an entire tape with a video signal that contains either time code or control track, which is necessary for editing. SMPTE time code, accepted as the standard by the Society of Motion Picture and Television Engineers, is the time code used most often, and if it is offered as a choice, you should take it. Preblacking could be compared to paving a road. Raw tapestock appears as snow on the screen and has no numbers of any kind. The machines can't read anything. If you choose to edit on this, it can be done, but you must use the simplest of the two basic modes of editing—assemble editing, or assemble—which is discussed later in this chapter.

Drop Frame and Non–Drop Frame Time Code

Video is said to run at the rate of thirty frames a second, but in reality, it's just over twenty-nine. When sticking to the actual time is critical, as in broadcasting, drop frame time code is used. Every thirty seconds or so a frame number is dropped to keep tape time consistent with natural time like leap year. Drop frame and non–drop frame are handled solely by the machines, and for demo tapes you don't need to worry about them. You can select whatever the facility or editor normally uses. In parlance, these terms are abbreviated to *drop* and *non–drop*.

Control Track versus Time Code

Another system for finding your place on a tape, control track, is quite different from time code. With control track, the machine counts pulses to tell you where you are. All the pulses are alike. When the tape comes to a stop, the machine can lose count and slip a few frames. Each time this happens, the tape gets farther away from the desired point. With time code, each frame and field is given a specific number, which is recorded right on the tape. In this case, when the tape stops and the machine loses count momentarily, it picks up again at exactly the right number.

Of special importance to us is the ability for the machine to find any spot on a tape with time code from any other spot on the tape. With control track, you must wait while the tape rewinds so you can zero the counter and go forward again to find your place. When you get bored with doing it that way, you start guessing where a scene is and then you lose your place entirely. Clock-time begins draining, and the conversation goes like this:

YOU: I'm sure it's right after the guy in the suit exits.
EDITOR: Which guy in which suit?
YOU: The mafia guy that comes back in and threatens me.
EDITOR: Have we seen that scene yet?
YOU: No, that's coming next. That's going to be later on.
EDITOR: So, do we want to go to the front of the tape or the back of the tape?
YOU: Back. We want to go back.
EDITOR: Okay. Now we're going to the end of the tape.
YOU (vocal pitch, up an octave or so): No! No! No! I want to go back. It's earlier than this.
EDITOR (vocal pitch up two octaves or so): Oh! You said the back of the tape. You mean the front of the tape.

This particular conversation usually concludes as follows:

YOU (screaming): There it is! There it is! You passed it! You passed it!
EDITOR: That's okay. We'll find it again. Uh, do you want to go forward or back?

DEMO TIP: If it's not upstream, it's downstream.

As with Niagara Falls, everything in video is either upstream or downstream. Upstream takes you to the beginning of the show. Downstream takes you to the end. When you get mixed up, and everyone does, think of Niagara Falls.

If you have logged your scenes carefully and are using time code, you will get to the desired scene the first time, every time. No one even has to look at the screen! You tell the editor the number, the editor types it in and presses a button, and the machine stops on the exact spot required. It's accurate to one-thirtieth of a second. Time code used to be recorded on one of the audio tracks, but now most machines use a special track known as the *address track*.

With time code each frame is given a specific address. Think of the difference between control track and address track as follows: Say you want to visit a friend who lives in the 375th house on Control Track Vista. To find your friend, you start at the first house and count up 375 identical little marks, one for each house. If you lose count, you have to go back to the very first house and start over. If, on the other hand, your friend lives at 288 Time Code Drive, you could turn onto the street at number 150 and just keep going until you get to 288. Like a house, each frame on your tape has its own address.

Unfortunately, many of your source tapes will not have time code, and your best defense will be to know them very well before you start to edit. If you log your tapes and then wait two months before you book the edit session, you will probably remember little of what you did. It is possible to post-stripe time code—that is, to record the time code after the recording is on the tape—on an unused or mixed audio track, but it's an extra expense and not worth it unless you have years worth of material.

Assemble Editing

Assemble editing preblacks the tape as you work and stops every time you stop. As mentioned in Chapter 7, you cannot separate the audio and video edits with assemble editing, and you run the risk of interrupted time code and hard edits, which are not smooth and make the viewer blink. If you are doing a straightforward job, such as a commercial reel or a collected works tape, or you're a news gatherer and you need to get your footage on the air in a few minutes, go ahead and assemble. But there are such important pitfalls to this that some editing controllers actually flash a warning light whenever the assemble mode is engaged.

If the editor has a momentary lapse of attention while working in the insert mode, there's little harm done. If, however, the editor, using the assemble mode, goes back to an earlier scene to make a little change upstream, and forgets to switch out of assemble, a portion of the tape will be erased and you will have to redo everything from that point on.

Assemble erases a small band on the tape just in front (downstream) of where you are working. Think of it as a snow plow on the front of a truck. You will need to push that pile of snow all the way to the end of the road because you cannot drive over it. In the same way, the snow field must be moved all the way to the end of the program to enable you to work on the tape. Picture cannot be inserted over snow. We've seen distraught people try and try and try.

This is one of the very few things that can go really wrong in an edit session, and it's not worth taking the chance. You might have just worked for six hours creating your tape in assemble, then decided to go back to add a title at the beginning. If the editor forgot to switch to insert (and every editor does, usually only once in a career), you will have to sit through those six hours all over again. You could make a submaster and lose a generation, but that's no fun either. Preblack the tape switch to insert and forget about the problem.

Many studios have preblacked tapes for sale—you cannot buy them at a store—and charge varying amounts for them. The more upscale the studio, the higher the price of the preblacked tape will be. A solution that works out fine for everyone is to preblack the tape at the beginning of the session during initial discussions of the project or while reviewing source tapes and laying out the edit plans.

This works especially well for demo tapes of less than ten minutes because you only need to black as much of the tape as you know you're going to use. If you buy broadcast stock to use for your edited master at a large tape outlet, just bring it to your session and ask your editor to black and code the amount of tape you'll need while you're getting started.

One small problem with this method is that you can overrun the amount you preblacked as you edit if you underestimated the program length. When this happens on a simpler system, you won't see the loss of sync until you play the tape back. This is why some editors justifiably insist on blacking a whole tape. Usually, however, the time code stops running on the status monitor when the black runs out, and the operator knows right away. All the operator needs to do is jam-sync more black and code, which picks up the exact minute, second, and frame where it left off, and you're back in business.

Insert Editing

This type of editing allows you to insert a scene between two others with a perfect in point and out point and to do a video-only insert, an audio-only insert, or any combination, all programmable, as long as the length of the new scene equals the length of the one being replaced.

DEMO TIP: Tape cannot be spliced.

Tape can only be edited by copying from one cassette to another, eliminating what you don't want, adding what you do, and copying scene by scene in the order you select. Insert editing allows you to go back and insert one thing *over* something else, but keep in mind that you cannot squeeze it in between or stretch it until it fits!

Suppose you have a scene where you pull up in a boat, then a scene where you walk up the dock, followed by a scene where you enter a beach house, and finally, a long scene with someone who's painting a seascape from the front window. As you edit the tape, you decide that the walk up the dock doesn't say much, so you cut it. The scene now plays like this. Pull up in boat, enter beach house, scene with painter. As you review the finished tape, you realize that the scene where you walk up the dock was important. Unfortunately, it's too late. You have two choices: Reedit from the point where you pull up in the boat all the way to the end or go to a submaster.

This is one important reason why those folks in mainstream television edit off-line. They can reedit all they like, for a couple of hundred dollars a day, until they have their show exactly as they want it. Then they head for their favorite on-line studio, where the price of changing your mind is $600 or $700 an hour. Your defense is to know as much as you can before you enter the edit suite and to apply the rules outlined in this book. If you do change your mind, try to change it as soon as possible. The sooner you change it the less you'll have to redo.

In the section on equipment later in this chapter, you will learn about the auto-assemble function, which can deal with this problem. Machines with auto-assemble remember every edit you made and can repeat the whole job perfectly, practically without human intervention.

Tracking

Tape is said to "track" around the drum where it contacts the video heads and tracking needs to be adjusted to maximize this contact. Each machine tracks a little differently. As a result, all but a few of the most recent consumer VCRs with digital tracking have a tracking adjustment. It's usually a little wheel in an inconvenient spot underneath the VCR, and it has an almost imperceptible little nick in the center where detente, the midpoint, should be. Try turning this wheel from one side to the other. You will notice that the picture breaks up at the top of the screen when you turn it to one side and at the bottom when you turn it to the other. As the machine's use increases, the tracking range narrows and shifts from the original detente position. The point where each tape tracks the best is halfway on the wheel between the start of breakup at the top of the screen and the start of breakup at the bottom and not necessarily at the little nick indicating detente. When you readjust the tracking wheel, you match the tracking of the machine the tape was recorded on to the one you're playing the tape on. Once they've had enough wear, no two machines will be alike. Eventually, you will not be able to get tapes to track at all. It's then time for a tape transport overhaul.

Tracking is exactly the same in an editing studio as with a consumer VCR—and much more important. Each tape you're going to edit from will track a little differently. A conscientious operator will adjust the tracking knob for each tape. When it gets busy, though, this is easy to forget. If you notice a crosshatch effect on the monitor or the same type of breakup as on your home television set, politely ask about it. The editor will set the tracking the same way that you did at home. This is done on the playback side only; the record tracking is automatic once the machine is set to record.

How Editing Works

Many who are unfamiliar with the editing process will be faced with trying to devise a way to tell their editor where to begin the edit. This can become a comedy, as people, in an effort to be polite, devise sentences of explanation about where to come in and sometimes even begin phrasing them as questions. Others will become so frustrated at not being able to make themselves understood that an emotion emerges that is not dissimilar to anger. This creates unnecessary friction and time-wasting misunderstandings. You want the edit made precisely and your operator only wants to comply. Here are some ideas that will clear the problem up once and for all.

Suppose you're searching for the spot on the edited master where you want to make your next edit. You find it: The plate of spaghetti connects with the dignified man in the tuxedo. "Boom!" you might yell. "I want to come in there!" Some people yell other things like "Now!" or "In!" but the more workaday among us snap our fingers. This is polite in an editing studio as long as it's not half an inch from the editor's ear.

We knew an extremely polite producer who believed that finger snapping was condescending to the operator. He used to bring along one of those little tin frogs that makes a clicking sound when you squeeze it. That was very thoughtful, but unfortunately the device made a double click. Were you supposed to come in on click one? Or click two? Or halfway between them? It was nerve racking. We prefer a middle distance finger snap and, barring that, a word we can recognize like "Boom!" One way to take the unpleasantness out of the finger snap is to aim it at the program monitor rather than the operator's head.

So "Boom!" we're going to a close-up of your look of shock when you realize who you've hit. The editor presses an in point on the record side, an in point on the source side (your close-up), an out point at the end of the shot on the source side, and the machines are ready to preroll.

The machines must run in synchronization (sync), and to accomplish this, a sync generator sends out a pulse to all the machines in the system, which it reads from the incoming AC wall plug, so that the machines all march in time and in step. They must be in perfect synchronization at the moment the edit is performed. The duration of the preroll will be anywhere from five to ten seconds. It must be six or more to ensure broadcast quality.

The videotape recorders (VTRs) back up their six seconds from the in points for a moment, then let go in a race to the finish. Source gets ahead, and record bumps itself to catch up; then source is behind, and it bumps but not enough, so record graciously brakes a tad, and by now they're two and a half seconds from the in point, and they've settled down into sync. During all this speeding up and slowing down, the picture wobbles all over the place, looking exactly like bad tracking. There probably isn't a soul among us who didn't call out at the first editing session, "That's not going to be on my tape like that, is it?"

Knowing Your "Ins" and "Outs"

In the following, as in an editing studio, the source machine and record machine will often be referred to as source "side" and record "side." We presume this is because these machines often sit on opposite sides of the edit controller. Synonyms abound in video production but usually their meanings are self-explanatory. We try to employ them all so you will get used to hearing them and be sure of their usage.

The *in point* is the place on the source tape or the record tape where you want your shot to begin. The *out point* is the place on the source tape or the record tape where you want your shot to end. The out point becomes the new in point at the moment the scene has been recorded. This is usually controlled by the record side, which automatically sets the new in point at the same frame as the old out point.

Let us explain with a little scene we call "Jerry and the Dog": Man throws stick; dog chases stick; stick lands in pond; dog skids to water's edge.

"Where should we come in?" asks the editor.

"Right where we went out," you say.

"I was thinking maybe we should up-cut that shot a little bit and go out before the stick hits the ground," the editor says. "If that's the new in point, we justify the dog's look skyward and his bark. That would pick up the pace."

"I was thinking just the opposite, actually. I thought we should extend the shot a little bit and see the stick bounce into the water. With that as the new in point, we can cut straight to the dog's skidding stop at the water's edge."

"That's a good idea too," the editor says. "Let's try it both ways and see which looks best."

"Maybe we can combine them," you reply. "Listen to this. We up-cut the old out point, as you suggest, and make the new in point where the stick is highest in the air. Then the new in point on the source can be where the dog looks up and barks. We go out right on the bark and cut to the stick bouncing into the water. That will be the new in. Take one second of that, approximately, and set the out there. Now, let's make the in for the source right where the dog digs in his paws and starts to skid. Whaddya think?"

There's no harm in going through it again; it's important. Take your time with the next couple of paragraphs. Try to imagine each step as you go along. The in point on the source tape and the record tape is where the shot begins. The place where you want the shot to end is the out point. The place on the record side where the shot has just ended will become the new in point. Although it is the place where you want to end the shot, for the editor to understand you flawlessly every time, think of it as the new in point on the record side. Understand that an out point can have nothing to do with where a scene is going to start. The new out points are now going to become where you choose to end the next edit. Think of it as something progressive: a pattern in motion. The out points can be chosen on either the source tape or the record tape but not both. The shot you've copied and the shot you've laid down are equal in length, so choose which side should determine the end of the edit, and the other side will follow automatically. This means there's one less point you have to worry about. Now there are only three things you need to control instead of four.

Let's go back to the spaghetti scene where you look shocked. It cut when you turned your head to keep from laughing. You and your editor chose that as the out point. The minute the shot was laid down, however, the machine automatically selected that last frame as the new in point for the next shot. The old out point is now history. For the new in point, you select the shot where your Saint Bernard, Churchill, scutters across the kitchen floor, leaps up on Dr. Rentwhistle (the president of the hospital where your character works), and starts licking off the spaghetti. Rentwhistle and Churchill stumble backward through the swinging door. As the door swings back and blocks the view, there is a terrible crash. Source in point: Churchill scuttling across floor; out point: crash. Record-side in point, set by the machine: head turn.

Preview and Perform

Because you've been using time code and you're certain about the new record-side in point, there's no need to waste time on a preview. Instead you perform the edit. Machines roll back, sync dance begins, the edit is made, the shot is laid down, the postroll of one second stops the machines exactly one second after the crash. The machine has set the new record-side in point perfectly, right at the crash. By editing like this, without making unnecessary previews, you can cut the time in half or better. If you make all possible cuts in half the time, you'll save half your bill on that aspect of the session.

Preview only for the following purposes:

- to adjust audio or video levels
- to up-cut the previous scene
- to see how the previous cut will look if you shorten it. (If you shortened it without a preview and didn't like it, you'd have to lay it down again.)
- to insert a shot between two scenes you will need to preview to be certain that all points are set correctly before you perform the edit
- to allow the editor breathing room. The editor is probably seeing your stuff for the first time and has a lot to think about. If the editor gets harried because you're pressing him or her to perform every possible edit, your session is going to slow up in other ways, and the fun will disappear.

Keep in mind that editors work at different speeds, and the slower ones may not be slower in the long run. Editors who pride themselves on driving a hundred miles an hour without looking back but leave flash frames or uneven audio on your tape, requiring you to repeat the session to fix all the problems, do not win the award for fastest editor overall.

More Technical Problems

Besides glitches, tracking, hot studios, and clogged heads, is there anything else to worry about in the technical department? Yes, a few. Let's start with sync.

Sync Lines

A sync line occurs when the machines are out of sync. If the machines have been miswired or if for another reason they are reading sync from different sources, a sync line will appear when you play back your tape. You will not see it when the picture is being laid down, just when you play it back. This is because of electronics to electronics (E to E). When machines are in record, the picture you see on the monitor does not come from the recording tape. It is fed from the source deck to the record deck and on to the monitor through the electronics only. This phenomenon is known as E to E. To see the actual picture on the tape, you have to wait until the scene has finished recording, then play back the tape.

If you notice a narrow horizontal line climbing slowly up the screen, there is a sync line on your tape. Once it disappears at the top it will reappear at the bottom. It can look like a physical crease because the picture is distorted, but it's not a crease, and no physical harm has come to your tape. Sync will have to be restored and the scene recorded again. You cannot tolerate a sync line because it will look far worse on the copies.

It is possible to get an entire copy that was recorded out of sync or to record a whole show off the air with the same problem. If this happens, get another copy. If you can't get another copy, you'll have to live with it. This will be one of those times that you'll wish you'd gone to medical school as your folks had wanted.

Flash Frames

DEMO TIP: If the tape makes you want to blink, review the edit.

A flash frame is not a major problem. It results if the edit came in one frame or half a frame late or if it came in early on the source side. You will see a flash of the scene you intended to cut. If the editor didn't see it, point it out, and the editor will trim the in point back a frame, redo the edit, and the problem will be solved. If you don't catch it at the time and if your source scene isn't long enough to insert over it, you may be in for some tricky reediting to get rid of it.

Flash frames occur less often with time code than without it, but they are part of the editing process, and you will be paying for the time. If flash frames occur often, look for the problems mentioned at the beginning of the chapter: that is, overheating, clogged heads or overly worn tape.

If you cut from a very bright shot to a very dark one, it will feel like a hard edit when nothing is technically wrong. You have two choices. You can select another shot, or you can do a slight fade to match the two brightness levels more closely. Either one will smooth out the edit.

Wrong Field Edits

Everyone notices when this editing problem occurs. It makes a big flash, looks sloppy, and takes your attention away from the scene. The technical term is *wrong field edit*, but this problem is also known as a *flag edit* or a *whip edit*. Every frame has two fields, and these fields interlace on the screen to make a single frame. Field 1 should always edit to field 2, and normally the machines take care of it automatically.

Under certain conditions, however, a wrong field edit can occur. If you have run out of control track and are inadvertently insert editing on raw, blank tape, every edit will whip wildly. If the room has gotten too hot, wrong field edits can occur, and if one of the tapes you're using is too old or the signal is low, the edit will be hard. This problem can also be caused by clogged heads or by a machine that needs maintenance.

The last way you can get a wrong field edit is by editing the first part of your tape on one set of machines in one establishment and finishing your tape at another. When machines are given routine maintenance, they are set to edit on one field or the other. If the two machines were set up differently, the first edit you make will whip. It will be the only one that does, but it will look nasty. There is a 50 percent chance of this happening, and it's no one's fault. The secret is to choose your studio carefully in the beginning and to stay with it if you can.

Sync Interruption to Peripheral Equipment

The last unsettling problem that could occur is not really a problem at all. If the monitor goes haywire and has two big lines drifting across it, one horizontal and one vertical, and there are ghosts of your former video drifting aimlessly across the screen, relax. A TBC has been switched off somewhere, generator lock (genlock) from a camera or computer has been interrupted, or house sync has been switched off. Let the editor find the error, and go back to work. Nothing has happened to your tape.

The Machines and What They Do

Each machine you will encounter is described in the following section starting upstream in the system, with the source or playback deck, and moving downstream until we get to the record machine. Downstream of absolutely all devices, however, is the program monitor. Because this is the monitor on which you will be watching your program as it is created, its description has been placed before the others.

The Program Monitor

This is usually the most expensive monitor in the system and can cost upwards of $5,000. It has extremely subtle adjustments, fine resolution, and excellent color. This is the monitor that answers the question, "Is that the color my tape is going to be?" Some studio engineers suggest there should be only one color monitor in the system, because every monitor produces slightly different colors, which can be confusing to the client, while the real test of color is the vectorscope. The program monitor can be adjusted to match the vectorscope and thus put all chroma doubts to rest.

Playback Deck

The playback deck can only play tapes and cannot record. This machine offers one great advantage in an editing suite: It cannot possibly erase a tape. The playback deck handles the source tapes during editing and when making copies. It is used for logging tapes and for watching video recordings on a monitor. Everyone calls it *playback*, and if you're a pro, you have to call it *playback* too.

Player-Recorder

The player-recorder is similar to the playback deck, but it has assemble record capabilities. It is often used as a playback deck in editing systems. It cannot be used as an editor.

Playback Monitor

The playback monitor is the monitor on which you view the source tape before it is edited. No editing occurs on this monitor, and your editing will not be seen on it. If you try to figure out what the operator is doing by watching this monitor, you will not succeed. That is why it is very important to locate the program monitor at the beginning of your session.

The playback monitor is used to search back and forth on the tapes for the scenes you want to edit. As a result, you will look at this monitor more than the program monitor. It is easy to get confused after a while, and if this happens, just ask again. Some simpler systems use only one monitor for both playback and program; it switches back and forth automatically. On a system like this, you know where to look right away.

Preview Monitors

There may be many other monitors in the edit bay, depending on the sophistication of the studio, and most of them are used for previews of some kind.

These might include graphics cameras, character generators, switcher effects monitors, and so on.

Status Monitors

The status monitors are the screens with the little numbers that tell the editor what is happening and where everything is going on the tape. If your editor hasn't moved a muscle for quite some time and you suspect them of nodding off, lean over and see if any of the little numbers are spinning away on the status monitor. If they are, everything is probably fine. If they aren't and nothing else is happening, get your editor a cup of coffee. He or she probably edited until 4 A.M. last night.

Engineering Monitors

The engineering monitors tell the editor all sorts of things about what's going on electronically and whether the video is meeting broadcast standards.

Waveform Monitor and Vectorscope

Technically, the waveform monitor and the vectorscope are the most important screens. They measure all of the various aspects of the video signal. See "Broadcast-Quality Video" in Chapter 15 for more information on these machines.

Time Base Corrector

The TBC is generally downstream of the playback monitor, which is downstream of the playback deck, so that the operator can see what the source tape looks like before it is corrected. The TBC is a miracle machine that corrects what would be called the "flutter and wow" on audio tapes, and other problems resulting from the instability of the tape. It does this by making millions of educated guesses about what a perfect output would be, and it digitally corrects the signal to conform to that. Once the TBC has the signal converted from analog to digital, it can be manipulated in all the magic ways you see on television every day. Digital time base correctors can stand alone and may be capable of digital video effects, or they can be built into the VTR and invisible to the eye.

Analog versus digital. Analog means "parallel," and it comes from the same root as *analogous*. An analog signal is a continuously varying signal, and its abilities are limited. Radios, radar, and television are all analog systems. They take a stream of electrons and bump them around from here to there, in tubes and transistors, until they've manipulated them into the behavior patterns that do the job. They do this by varying the voltage. In television, the maximum is one volt.

Digital is related to digits, or numbers. Computers work by making the electron decide whether it is a one or a zero, an on or an off. When you digitize any numeral in a computer, you convert it to a string of ones and zeros. Once the computer, or TBC, which is simply a computer dedicated to a specific task, has digitized the analog signal by taking periodic measurements of the analog signal and recorded these measurements as binary numbers, you can now manipulate this digitized signal any way you want with no degradation of signal and no generation loss. Once you're working in ones and zeros instead of amplitudes, you can paint a mustache on the Mona Lisa.

The reason that video and computers have remained incompatible so long is that television is still dependent on an analog system. To make the editing system work, we convert the analog signal to a digital one, manipulate it, then convert it back to analog before sending it downstream to the television.

Switcher

The switcher, or special effects generator (SEG), is the machine that handled the effects before the DVE came along. It is still a necessary machine that performs many functions. It connects all peripheral devices to a central box that has little illuminated buttons all over it. The peripheral devices include cameras, VTRs (for A/B roll, especially), character generators, graphics computers, and often audio. When the switcher has the "audio follow video" feature, you can dissolve audio with the video automatically.

The switcher creates the dissolves in which one scene cross fades through another. The switcher also creates the wipe patterns in which one scene is wiped off the screen as another replaces it. Wipes can occur horizontally, vertically, or in an astonishing number of other patterns, with a border or without.

Finally, the switcher also permits the familiar chroma key effect. The switcher cuts a key window wherever it reads a particular color. That is, it makes a hole in the picture so that whatever is behind it can show through. Blue is the standard color (although any color would do), and now actors always know not to wear a blue shirt to a chroma key job.

For this effect, you roll two tapes. On one tape, you're in front of a blue wall saying your lines. On the other tape, you have a rain forest. Wherever the switcher reads blue, it shows the rain forest, and wherever it reads any other color, it shows you. If you fibbed about your eye color to your agent because you needed the job, the audience will see a rain forest right inside your head.

The process of keying logos was discussed in Chapter 6. This is far simpler and less time-consuming than designing a logo on a computer, and it can be even faster than using a character generator. Some studios have an add-on charge for the Chyron. If this is the case, use the switcher to add borders around your keyed letters, and as long as your artwork is good, no one will think twice about it, once the tape is made.

Graphics Camera

A graphics camera can be any video camera used for recording artwork. The facility may have one, two, or more cameras hooked into the system. Some of the older cameras, such as the JVC KY1900, make excellent graphics cameras even by today's standards. Their main drawback is that they require a substantial amount of light. Also, they are tube cameras, and the tube can develop a burn-in if the camera is left pointed at a bright spot of light, without being capped conscientiously.

With all cameras, check the picture quality on the screen. If there is more than one camera, note how well they are matched. A photo montage can look amateurish if the two cameras used were obviously very different.

Character Generator

Character generators, which create titles, vary widely in the quality and the styles of font they are able to generate. Use the highest quality CG you can line up; the better it is, the more professional your tape will look. As mentioned in Chapter 5, Chyron is one of the finest titlers and is used by the networks. People often refer to the CG as a *Chyron* even when it's made by another manufacturer.

Digital Video Effects

More and more switchers have DVEs built right into them. A studio will charge for digital video effects according to the cost of the machine they purchased, and the cost of these machines varies quite a bit. Panasonic builds a DVE that sells for $2500, and Quantel sells theirs for up to $80,000. An especially elaborate machine, known as the Ampex Digital Optics (ADO) because Ampex invented it, is for those who are very serious about their video. But there are all sorts of DVEs in between. The effects created by Video Toaster, which was discussed in Chapter 5, are all digital video effects.

Depending on their sophistication, these machines can take your moving video image and twist it, squeeze it, zoom it, tumble it, spin it, make it fat, make it skinny, or turn it into an hourglass. They can make it sparkle or pulsate, gleam like a star, or shrink into a black hole. To achieve these effects, you push the correct buttons and then manipulate the joystick, mouse, or trackball. Unlike computer graphics, where the more sophisticated images must be recorded frame by frame, this all happens in real time.

Audio Mixer

The audio mixer takes audio from various playback machines, including the compact disk player, cassette player, reel-to-reel tape, VTR, turntable, and microphone. It measures the audio levels on your tape with volume unit (VU) meters, which measure the amplitude or volume of sound. As your audio is recording, the VU meter should peak at approximately zero, not fluctuate into the red too often, and not "pin"—that is, touch the little pin at the top of the range that stops the needle from going too far. If the needle rests against the pin, even momentarily, the audio will be distorted, and you will have to redo it.

Some video editors treat the audio like an afterthought, and they pay less attention to consistent audio levels than is required. There's nothing more annoying than a tape with audio that blares out at you one minute and that you must strain to hear the next.

Edit Controller

The edit controller makes everything happen. All of the tape recorders are connected to it. Some controllers can issue commands to the switcher, and some have the ability to perform certain effects on their own. The controller is a dedicated computer—that is, a computer designed to perform specific functions. The more sophisticated this machine is, the faster your editing session will go. Many of the things it can do are beyond the needs of the average demo tape producer, but it helps to be aware of them.

For example, the Convergence 195LM controller is a mid-range A/B roll machine that is used in a lot of houses that work at the industrial level. It can control two source machines and one record machine, enabling you to do A/B roll and tape-to-tape effects. It also has an internal list management system, which is extremely useful because it remembers all the edits that you make until you turn it off. If you decide to change any edit, you simply scroll through the list using the joystick until you reach the edit you want to change. The machine will quickly set everything up for you, saving you as much as half an hour on the clock. Although the list is lost when the machine is turned off, it will output its list to one of many list management programs that run on PCs and import a list as well.

With the auto-assemble feature, the controller is capable of assembling every edit in the list automatically. This is useful when you want to remove a particular edit, for example. All you need to do is to go back to that edit, eliminate it from the list, and press auto-assemble. The machine will redo every edit for you from that point on.

The scene store is another useful feature. Suppose you want to try an edit in a variety of different ways. The operator can store all of your options (up to 99) on a scene store list, and you can try them one at a time. You can go back and forth, and the machine will reset each one with a couple of key strokes. Key strokes relate to editing speed: The fewer the key strokes required, the faster you will go.

With the edit controller, you can program your switcher effect by frame number and trigger the exact effect and its duration, rather than doing it manually. You get exactly what you order the very first time. If you want to alter the effect, it's as easy as trimming a few numbers.

Remember that everything you need to do twice doubles your cost. The more things you can program into the edit controller, the shorter and cheaper your session will be. When comparing systems, balance the sophistication of the machine against the hourly rate. Generally, the more it costs, the more it can do.

We used the Convergence 195LM controller simply as an illustration. All of the major companies build excellent edit controllers. An edit controller has recently become available from RGB Industries for the Amiga computer, and although few of them are around currently, they will soon become available. When combined with the Video Toaster, the results should be excellent. An extremely sophisticated PC-based computer, the Case, is supplied by ASC Video Corp. of Burbank, California. Because of its versatility, even at the highest broadcast levels, its modest price, and its ease of use, it is quickly becoming the machine of choice with many editors.

Editing Videotape Recorder

The editing videotape recorder is located downstream of everything except the program monitor. The tape is actually made in the editing VTR, which is capable of insert and assemble recording.

Protecting Your Tape from Accidental Erasure

At the end of a lengthy, intense editing session sometimes lasting six hours, sometimes lasting twenty, everyone is exhausted and a bit punchy. This is the moment when accidents happen, and this is the most likely time that an erasure will occur. Let us, then, end this entire section by repeating the following advice.

DEMO TIP: Eject your edited master and remove the red button.

Part IV

Distribution

18 What to Do Once Your Tape Is Made

You can have the world's finest demo tape, displaying Academy Award–quality talent, but if you don't show it to anyone you won't get hired.

You must make copies, package them well enough to reflect the excellent material on the tape, and then get them out. Remember the enthusiasm you expressed when you watched your whole tape for the first time in the editing studio and you couldn't help applauding at the end? Keep that feeling alive when you make your copies. If you are doing a general mailing to agencies, keep it alive when you send your head shot that includes a resume containing the words, "Tape Available on Request." When you're invited into an agency to show the tape remember how good you feel about it.

If you're someone who's made a tape for a product, service, or invention, follow the same advice. Even if you're a low-key person, be on fire about it. Your enthusiasm will be infectious. The tape you made is a tool. Now it's time to put it to work.

Making Copies

Some duplicating places allow you to supply your own tapestock. You can get great deals on tape if you buy it in bulk. You may want to get together with friends because the more you buy the cheaper the tapes are.

Whatever the purpose of your tape, decide how many copies you will realistically need and how many you can afford. Remember, you can make more copies at a moment's notice. If you're an actor, did you target three specific agents, or did you pick up the SAG agent list and decide to start with the A's? Some people start with the Z's. Does anyone ever start with the N's or the K's?

If you're a businessperson making a career change, get an idea of the number of companies that will decide to view your tape, then run your copies accordingly.

If you require only a few to start with, copying the master where you edited your tape will be the most convenient, even if it costs a little more. If you decide to use tapes you already own for copies, follow the advice in the packaging section. Once the master is completed and in your possession, a duplicating house likely has the best copying deal.

The quality of bulk tapes can be unacceptably low. Buy a few tapes first, and check them out. Bulk outlets are often located in industrial parks and should seem well established. Ask for high-grade tape and find out with what brand of stock they load their cassettes. Listen for a name you recognize. Then make the following evaluations. Does the cassette feel cheap? Does it rattle alarmingly when you shake it? When you record on it, are the colors all smeary? Does it jam in your machine, or even come apart completely? If everything checks out fine, you've found a place to save on tape. There are also excellent mail order places and they can be found by looking through video magazines such as *Videography*. If you can afford to pay a higher price for your tape, we recommend buying the highest grade you can from a discount store and not taking any chances. The other drawback to providing your own tapestock is that duplicators usually charge a fee for "client-provided stock," and this fee could wipe out your savings.

When choosing a duplicator, it's important to make sure that the machines used to record are professional, industrial machines, rather than consumer machines. The cost per copy will be a bit more, but the difference in picture quality will be worth it.

Packaging

Think some more like a casting agent at the office. Can you tell an actor by the cover of the tape? When all those folks are sitting around and someone picks up the next tape, imagine that it's yours and the cover is clean, clearly typed, and labeled on the front and on the spine. Now imagine this. The tape they pick up is yours, only this time it's been revised several times, and you've scratched out the list of scenes with different colors of ink and some pencil. You've put asterisks next to certain scenes and scribbled in some illegible counter numbers below. The label is smudged here and there by a dirty eraser. You've written in the name of your new agent but failed to completely remove the old one. The plastic sleeve has split in two or three places, so you repaired it with transparent tape, except for the bottom corner where you used some masking tape. The whole box is gummy to the touch. Remember the line,

"Your tape is your audition?" Well then, the box of your tape is like a first impression. It's you coming through the door.

Now think about this: The cassette box is free advertising space. Make use of it by putting your head shot on the box. There are several ways to do this. A little picture on the spine above your name works well because the tapes are often stored with the spine facing out. Every time agents and casting agents reach for a tape, they see your face matched to your name. When you buy your library cases, which fold open like a book, at the discount video store, ask for full sleeves that cover the whole box. You can reduce your photo to the size of the cover and insert in the sleeve. You can put a different head shot on the back or use the back to list the tape contents with the running times. The tape contents can also go on the inside left-hand cover with addresses and phone numbers.

Agent John Hugh came up with a great idea. He cuts down a standard 8″ x 10″ photograph to fit a cassette cover, and it turns into a great choker of a close-up! It's very effective and practically free. You can also cut your name off the same shot and add it to the label. Whenever you need to change the information on a label, change the whole label. Let it be clean, easy to read, and professional.

For business video resumes, a straightforward, clean label with your name imprinted on the front and spine makes the best impression. You might consider placing your mailing address and phone number in the lower left- or right-hand corner. As with actors, including a S.A.S.E. with your tape gives the impression of politeness and professionalism.

For a product demo, the label can be as elaborate as an advertisement, and you can use your imagination to help make the sale.

For all who will be mailing tapes, we are told that videotapes are not affected by X rays and we have never seen any evidence to the contrary. Padded mailers work just fine for sending tapes, although we always recommend library covers for cassettes.

Finding an Agent

If you are looking for an agent, let prospective agents know that you have a tape that you're very happy with (assuming that's true) and ask them if they would be willing to view it. If they agree, send a copy of your tape along with your photo and resume and include a self-addressed, stamped envelope (SASE) to make it as simple as possible for them to return your tape.

There is a theory circulating that if you don't send a SASE, agents will be less likely to reject you because then they will have your tape lying around and may be driven to phone you to come and pick it up. This gives you the opportunity to harass them into taking you on. Next to talent, the most important factor in getting work is having a good rapport with your agent. Trying the above theory is not the way to initiate a good rapport.

Agents will sometimes schedule a meeting so you can view the tape together. If this happens and you are given a choice of times, pick later in the day because the business day is winding down and you will be less likely to be interrupted. If the agent has to take phone calls when you're viewing the tape, it can be very distracting. Should the agent miss an important scene, there's no harm in politely asking to back up the tape.

There are many small five-inch player-monitors on the market today that are excellent for desktop viewing. If you have one of these, you might bring it along, just in case. When viewing your own machine, you can pause it when there is a distraction and start it up again when attention is again focused on you.

Distribution Don'ts

Some actors have gotten work with the "scattershooting" method of sending out tapes. They have their tape duplicated in bulk and send the copies out to everyone they can think of. It is an expensive method, and it risks a good relationship with people you will meet down the line. Los Angeles is an astonishingly small town sometimes. If you annoy people so much that their bodyguards phone you instead of their secretaries (we know a person this happened to), it's unlikely you'll be working for them in the near future.

There are some approaches you should never attempt under any circumstances, like hand-delivering your tape to a producer's home. Not only will you have to look out for the bodyguards, but there'll also be the security guards, the guard dogs, the live-in maid, and probably Neighborhood Watch. Whether you make it back to Sunset Boulevard will depend on the length of your stride and the intensity of your terror.

We recommend that you seek an audience rather than lobbing your tape through a window. We know the technique worked on Brutus, but the guy who chucked a note at Napoleon got into big trouble.

International Standards Conversion

The United States introduced the first television system in the world, demonstrating it to the general public at the 1939 World's Fair in Chicago. Around that time, the National Television Standards Commission (NTSC) was formed. The NTSC set the standard for television resolution in the U.S. at 525 lines per screen. This standard has not changed, and it is used in the United States, Canada, Japan, and elsewhere.

By the time the British got their system going, technology had improved quite a bit. They developed their own system, phase alternate line (PAL), based on a standard of 625 lines. PAL is used today in Britain, Australia, New Zealand, and other countries.

The French also invented a system of their own, séquential couleur à mémoire (SECAM), which they sold successfully to other countries.

When you need to send a tape overseas or if you have just received one from overseas, you will call a place that advertises international standards conversion. Tell them the origin of the tape, and they will transfer the tape to the appropriate standard.

Prices at these establishments vary widely for two reasons. First, there is a broad range in the quality of the machines they use. International standards conversion is an expensive process in any case because it is so specialized. Second, the high-end transfer houses charge higher prices because they have higher overhead costs across the board. If you are converting a copy to go to a relative overseas, try to get a deal, but if your tape is going to Network 9 in Sydney for national broadcast, don't take any chances.

There are no differences in tape formats, by the way. VHS is VHS everywhere, and the same goes for all the other tape formats.

19 The Future

The future of promotional video is tied to the evolution of the medium. IBM has signed a pact to share technology with Apple. They will drop the Intel chip in favor of a tiny monster from Motorola that will revolutionize computing—again. Video is making a beeline toward digital at all levels, and this means that video and computer technology will soon merge completely. Desktop video is with us now, and with Hi8, S-VHS, and Video Toaster, all those unattainable effects that cost thousands of dollars just a few years ago will drop right into our price range.

CD ROM disks—prerecorded optical disks—containing whole encyclopedias are currently on sale. Sony has established a software company to make more. A firm in New York is already distributing to studios and producers a CD ROM disk that contains cinematographers' reels. If you phone the firm and send them your reel, they'll add it to the disk.

We will soon be buying re-recordable compact disks that will put an end to videotape. Panasonic already has one of these "rewritable" CDs on the market. Soon you'll be "taping" your shows on a tiny disk. There'll be no more shuttling back and forth looking for scenes and no more degraded picture quality.

Digital audio tape (DAT) recording is in use everywhere. The end of generation loss has finally arrived. With this new technology, the millionth copy will be identical to the first.

Everyone is pushing hard for high-definition television (HDTV), but the U.S. government is trying instead to implement an improved standard screen for general consumption. Some studios have invested heavily in HDTV, however, and you'll soon be able to buy an HDTV set to play your videos on at least. Another new item is the flat-screen television set people have been awaiting, and the tiny video player with its liquid crystal display screen is a harbinger of this approaching technology. As computers and video merge, your entertainment center will become your communication center, and you'll be able to send information as well as to receive it. If someone needs your demo reel, video-fax it over.

What will not change? Good old three-quarter-inch tape will be with us for a while yet because all our archival shows are stored on it. No one will want to transfer all that footage to something else unless there's a very good reason.

Editing will not change. It doesn't matter how the machines are configured or how big or small they are. It will still take talent, skill, and a lot of experience to know how to edit, whether your material is stored on film, videotape, or a twenty-five-cent piece. The machines can do a lot for you, but the decisions of what, where, and for how long will still be solely yours.

No matter what else becomes digital, our lives will always be analog. We live them in real time. So we'll be doing tapes or disks or chips, and we'll be hiring professionals to help us. We may not need them as often—there'll be more we can do at home—but some people will get real good at shooting and editing video, and we'll be asking them to help us make our reels.

Glossary

A/B roll: A video system that incorporates two source machines and permits dissolves, wipes, and other transitions between two moving source tapes.

add-on: An extra charge for the use of additional machines in an editing suite, such as a character generator or digital video effects machine.

address track: A track on videotape, separate from the audio and video tracks, that is reserved for time code.

ADO: See *Ampex Digital Optics*.

air check: A show that has been recorded off the air for later review.

Amiga: A computer created especially for reasonably priced video effects, titles, and graphics.

Ampex Digital Optics (ADO): Extremely sophisticated digital video effects created by the Ampex Corporation for use in broadcast studios. Pictures can be squeezed, wrapped, mapped, zoomed, and manipulated in a variety of ways, mostly in real time using switcherlike black boxes.

analog: A continuously variable voltage in an electronic circuit that can be copied, creating a parallel electronic image of the original. See also *digital*.

assemble editing: The recording of new video on a blank or previously recorded tape. This method erases the tape in advance of the recording heads as it is recording, and lays down a new video signal. In this method, each scene must be laid down in sequence, and video, audio, and control tracks are recorded simultaneously. It is the simplest form of editing.

audio blip: One or two frames of stray or leftover audio that make a blip or popping sound.

audio gain: The amount that the audio signal is amplified (volume).

audio-only insert: The recording of audio on a tape without disturbing the video. This can be done on audio track 1, audio track 2, or both tracks simultaneously.

auto-assemble: After programming a series of edits all at once into the edit controller, the computer can then assemble all the scenes without further input from the operator.

back light: A light placed behind the subject to create a sense of separation from the background. This also puts a pleasing shine in the hair and an edge of light on the shoulders.

barn door: A square panel attached to a lighting instrument to eliminate light spill from unwanted areas. Each instrument usually has four barn doors, although cheaper sets come in pairs.

Beta: Refers to both Betacam and Betamax tape formats.

Betacam: An in-the-field recording system that uses an expensive professional camera and a special high-speed half-inch cassette tape format. Beta systems exist to handle this format in postproduction.

Betamax: An early consumer tape format that employed half-inch tape contained in a cassette.

black: A tape that has a synchronous video signal and control track before editing or recording. It may also have time code. Black also refers to 7.5 IRE whatever its source, such as a black generator.

black level: The lowest level of the picture signal. It should not be set below 7.5 IRE. Sometimes referred to as *setup* on time base correctors and processing amplifiers.

bloom: The washed-out area that occurs when the luminance has peaked above 100 IRE in a video picture.

broadcast quality: A nebulous phrase that suggests that the resulting video would be acceptable for broadcast. True broadcast quality means that the technical aspects of the signal meet Federal Communications Commission standards.

B-stock: Equipment that is considered good enough to be sold as new but has been out of the box and probably used.

bump: The automatic speeding up and slowing down of machines during preroll in order to synchronize all tapes for the edit.

bump up: To transfer a lower level videotape format, such as VHS, to a higher one, such as three-quarter-inch or one-inch tape.

burn-in: (1) Visible time code numbers inserted in an off-line videotape for the purpose of making edit decisions. Often referred to as *window code* or *window dubs*. (2) A mark created on a camera pickup tube that may be temporary or permanent, caused by an excessively bright object or extreme contrast.

cable-ready tuner: A receiver in a VCR that can operate without needing a cable box to switch channels.

calling card tape: A very short tape that serves to introduce a performer. It sometimes consists only of a montage.

camera control unit (CCU): A black box used to keep cameras matched in a multicamera shoot. It can also control some camera functions remotely.

camera edit: Stopping the camera to change the shot rather than making the cut in editing. This can only be done with cameras built to make synchronous ("glitch-free") edits.

camera plate: A flat piece on a tripod to which the camera is attached.

cassette elevator: The door and mechanism that receives the tape in a VCR.

CCD: See *charged coupled device*.

CCU: See *camera control unit*.

character generator: An electronic machine used for generating titles on video.

charged coupled device (CCD): A chip that replaces the tube or tubes in a video camera.

chroma key: The process of cutting a key window in a video picture by electronically making one color transparent.

chrominance: The amount of color saturation in a video signal.

Chyron: A character generator enjoying wide use in the industry.

clip: A short scene taken from a movie or TV program.

collected works: See *compilation reel*.

color bars: A color chart used for adjusting a video signal.

color correction: The adjustment of incorrectly recorded colors on a videotape.

color shift: When the colors begin to slip out of correct adjustment. All colors shift the same amount, requiring hue or tint adjustment, which is technically known as *subcarrier phase control*.

color temperature: The number of degrees Kelvin used to describe different colors of light read by a camera. The higher the temperature, the bluer the light; the lower the temperature, the redder the light.

commercial reel: A collection of a performer's commercials.

compilation reel: A tape containing all of a performer's material assembled for later use in making a demo reel. Also known as *collected works*.

consumer camera: A smaller camera made especially for home or amateur recording.

contrast: The difference between the brightness and black levels in a video picture.

controller: See *edit controller*.

control track: Pulses recorded on a tape during the primary recording of a video signal to allow the machine to find and mark the tape's position during recording or playback.

copy: The act of transferring the contents of one videotape to another without editing. See also *dub*, *dupe*, and *transfer*.

cross fade: To slowly bring in one video signal, one audio signal, or both, while simultaneously fading out the previous signal.

cut: The simple electronic splice of one scene onto another.

cuts only: A simple video editing system that uses only one source machine and one record machine operated by an edit controller.

deck: A videotape recorder or player.

demo reel: A promotional tape created for the purpose of seeking employment or promoting a product or service.

detente: The optimum midpoint of an adjustment; particularly useful when adjusting tracking. Tracking detente is halfway between picture breakup turning clockwise and picture breakup turning counterclockwise.

diffusion screen: One of various screens and materials used to soften the light falling on a subject.

digital: The conversion of an analog signal to binary digits used in computers. The signal can be manipulated or enhanced before being converted back to analog for video or audio recording or display. See also *analog*.

digital tracking: An automatic system of digitally correcting minor tracking errors.

digital video effects (DVE): A variety of special effects that can be produced by manipulating the digital signal, such as posterization, mosaic, and strobe.

director's copy: Usually a three-quarter-inch archival copy of a show that can be borrowed from the production office. A misnomer because the director usually has a personal copy.

dissolve: An effect used to cross fade one picture into another.

distortion: Audio that is recorded too loudly and loses its definition or breaks up.

dropouts: Loose or lost oxide particles. They cause a momentary loss of radio frequency signal and are most noticeable at the heads and tails of videotapes. They appear as bright pinpoint flashes. Dropouts increase with the tape's age, the number of passes—times it's been played—and the conditions under which the tape has been stored.

dub: To make one copy or very few copies of an original. See also *copy, dupe,* and *transfer*.

dupe: Short for *duplicate,* to make many copies of an original.

DVE: See *digital video effects*.

edge damage: A narrow snow field across the bottom or the top of the screen. It is caused by playing a tape in a source or record machine in which the tape transport is not in good working condition.

edit controller: A computer that controls the operation of the source and record machines as well as the entire editing process. It often controls the switcher and other peripherals, such as the edit list management computer. Also known as an *editor*.

edit decision list: A list of the desired beginning and end points for the video and audio for each scene. This list can be written out by hand or created automatically by the edit controller or other computer. It contains other information such as tape or reel number, type of transition, and which video and audio tracks are involved in the edit.

edited master: The tape on which the program is created. It will be used for making copies.

editing system: A grouping of machines used to perform the various functions of the editing process.

editor: The operator or person directly responsible for the physical creation of the edited master tape. An editor may simply give instructions to an operator in a broadcast situation. See also *edit controller*.

effects (F/X): The special effects that the various machines in an editing system are capable of producing, such as freeze frames, wipe patterns, keys, and other manipulations of the video signal.

electronics to electronics (E to E): When the radio frequency video signal passes from the source machine to the record machine and out to the monitor without showing the actual recording as it will play back from the record tape.

ELP: See *extended long play*.

EP: See *extended play*.

E to E: See *electronics to electronics*.

extended long play (ELP): The slowest speed, usually six hours, on consumer VCRs. It cannot be successfully edited and is useful only for home recording.

extended play (EP): The slow speed on consumer VCRs. When there are three speeds, extended play often refers to the slowest of the three, the six-hour speed. The middle, or four-hour, speed is referred to as *long play*. See also *extended long play*.

fade: A dissolve from black up to picture or from picture down to black. Fades to white are becoming increasingly popular as midprogram transitions.

fill light: The softest of the three lights in a three-point lighting system. It is used to illuminate the subject on the opposite side of the key light.

film-style shoot: Using a single camera and reshooting a scene from different angles for final assembly in postproduction.

flag edit: See *wrong field edit*.

flash frame: Occurs when the new edit point enters a field or a frame late, leaving a flash of the unwanted picture.

fluid-head tripod: A tripod head that is filled with an oil in order to make all movements, particularly starts and stops, smooth and unnoticeable.

flying erase heads: An extra set of heads in camcorders and VCRs that helps make possible glitch-free edits. They are also responsible for the availability of playback effects such as still frame and slow motion, neither of which can be copied successfully.

footage counter: Numbers on a VCR used to keep track of the location of scenes when time code is not available.

formats: The various sizes and types of videotape and video recording systems, such as Beta, MII, or one-inch type C.

frame: A single video picture made up of two interlaced fields.

freeze frame: A digitally stored image of a single frame that can be recorded or manipulated. Not to be confused with a still frame created by pausing a VCR; a still frame cannot be successfully recorded.

Freon: A chemical used for cleaning electronic equipment, especially the heads and tape transport of VCRs.

F/X: See *effects*.

gain: The amplitude of a video or audio signal. With audio, the higher the gain, the louder the sound. With video, the higher the gain, the brighter the picture. Some cameras have an artificial-gain switch, which allows the brightness to be increased electronically for shooting in low light.

generation loss: The continual degradation of picture quality that occurs with each copy made of a copy. Digital copying will eventually eliminate this problem.

glitch: A nontechnical term used to describe picture breakup from a variety of causes. In an edit bay, it often refers to physical tape damage.

graphics camera: A camera, often mounted on a stand or pointed at a board, used to shoot stills and graphics.

gray scale: A measurement used to determine the correct gradations of contrast when setting up cameras and video equipment.

HDTV: See *high-definition television*.

Hi8: An 8-millimeter video format that permits broadcast-quality results from a consumer-size camcorder.

high-definition television (HDTV): A new high resolution television system. Several versions are under consideration for future television transmission. One version uses 1125 lines of resolution as opposed to today's U.S. standard of 525. The aspect ratio of HDTV is 16:9 rather than 4:3, making the television screen resemble the shape of a movie screen.

high grade: videotape with improved signal-to-noise ratio, lower dropout levels, improved binder, etc., which provides reduced color smear, brighter colors, and sharper images than standard tape.

hot spot: See *bloom*.

hue: The tint of the video picture that occurs across the entire color spectrum. It must be adjusted when faces appear to be the wrong color. If white balance was incorrectly set when recording, the results cannot be corrected without a sophisticated color corrector.

in point: The point on the source side and the point on the record side where the edit is set to begin.

insert editing: Inserting a scene between two existing scenes. The in point and the out point will be clean (synchronous). A control track must exist on the tape before insert editing can be done.

interlace: The two fields of scan lines that make up each frame of video. These two fields are alternately scanned, interlaced. This reduces flicker in a television picture.

international standards conversion: The process of transferring videotape from one standard—NTSC, PAL, or SECAM—to another.

IRE: The scale used to measure video amplitude (brightness) on a waveform monitor, from the acronym for the International Society of Recording Engineers.

ISO shoot: To do a multicamera shoot using separate (isolated) cameras and recorders. The resulting angles are edited during postproduction.

jam sync: A method of laying down time code during editing, which permits stopping and starting of the tape. The time code numbers begin recording from the last number laid down on the record tape. It is useful when assemble editing a tape that will later require insert editing.

jog: A set speed, usually faster than normal, at which tapes can be scanned.

jog shuttle: A wheel on many videotape recorders, including consumer versions, that permits scanning of the tape at varying speeds, from still frame to many times normal speed. On some edit controllers, a joystick handles this function.

joystick: A controlling lever on an edit controller that is used to move tapes at variable speeds, from still frame to many times normal speed. It is also used to switch control between source and record machines during editing.

jump cut: A jarring cut that alerts the viewer that time has elapsed and a section of video has been removed from an otherwise continuous shot. Although an interesting device when used in music videos, it usually indicates careless editing.

key: To superimpose a title, graphic, or actor over another picture, separate background, scene, etc.

key light: The primary and strongest light in a lighting setup. It is set slightly off to one side of the camera and a little higher than the subject.

key window: An area of a video picture that is electronically eliminated so it can be filled by a second video source, such as a title or matte.

lavalier: A tiny microphone with an omnidirectional pattern that is pinned to the subject's clothes. It can be hard-wired or radio-transmitted.

list management: A computer program designed to store and manipulate an edit decision list.

log sheet: A sheet used for manually keeping track of edit in and out points. It includes space for information about the chosen scenes.

luminance: The degree of brightness in a video picture.

master tape: The tape that contains the program from which copies will be struck (made).

modeled lighting: The act of creating a balanced lighting plot to create definition and the illusion of three dimensions on a subject's face.

moiré (mwah'ray): A rainbow effect caused by tight patterns, such as herringbone tweed, that confuse the electronic sensors in a video picture. Also, a condition similar to the stage coach wheels that appear to turn backward in old Westerns and which is due to the frame rate of film cameras.

montage: A series of cuts, not necessarily related, that create an overall impression with the viewer.

movie of the week (MOW): A generic term for any one-time, television fiction presentation lasting over an hour.

MOW: See *movie of the week*.

MII: A half-inch tape format used in broadcast applications.

multicamera shoot: A play or event taped by using more than one video or television camera recording simultaneously.

new-in-the-box: A piece of electronic equipment that is brand new and has never been opened.

NTSC: The 525-line standard for television resolution, which was adopted by the National Television Standards Commission for broadcasting in the U.S.

off-air recording: Recording a television or cable show with a videotape recorder.

off-air source tape: A tape on which off-air material has been recorded.

off-line editing: The process of assembling a show on a cheaper system in order to make editing decisions before going on-line to cut the final product.

omnidirectional mike: A microphone whose recording pattern is nearly 360 degrees and which picks up most sounds from a given location.

on-board recorder: A videotape recorder connected to the back of a camera. It can be permanent or removable.

one-inch tape: The videotape and format that is presently the broadcasting standard. It is slowly being replaced by smaller formats.

one pass: A tape that has been recorded on only once.

on-line editing: The process of making a tape that will be the finished product. It is not related to the sophistication of the system used.

operator: An editor or another person who actually operates the editing system.

out point: The point where the scene currently being recorded is set to end.

oxide: A ferrous (iron-containing) substance made up of minute particles glued to the videotape by a binder, upon which video images are magnetically stored.

PAL: The British, German, and Australian television standard, from the acronym for *phase alternate line*. Used in many Commonwealth countries, with the exception of Canada, whose standard is NTSC. Its maximum resolution is 625 lines.

perform (an edit): To record an edit, as opposed to previewing or replaying the edit.

pin: Audio that has been recorded at too loud a level and registers beyond the limits of the meters is said to "pin" and will be distorted during playback. It must be re-recorded at the correct volume.

pinch roller: A mechanism inside a VTR that holds the tape against the capstan (the spindle that actually moves the tape along at a constant speed). A worn pinch roller can be the cause of a number of tape transport problems.

playback: Playing or watching a previously recorded tape or scene.

playback deck: The videotape player that plays the source tape in an editing system.

playback monitor: The monitor on which the picture from the playback deck appears.

playback only: A videotape player that has no recording capabilities.

player only: See *playback only*.

player-recorder: A machine that can play tapes and record tapes but has no editing capabilities.

post: An abbreviation for *postproduction*.

postproduction: The final stage of a production in which the program is edited, audio added and sweetened, and titles and graphics included.

preblacking: Laying down on a tape a video signal that includes either control track or time code. Black is recorded at 7.5 IRE.

preroll: The roll back of two or more machines for a specified number of seconds in order to match up their forward movement, resulting in a synchronous (clean) edit at the edit point.

presentation reel: See *demo reel*.

preview: To view an edit or series of edits without actually recording the material.

preview monitor: A monitor used to view what will eventually appear on the program monitor. Titles are usually constructed on a preview monitor.

professional grade: The highest grade of VHS tape sold in tape stores. Sometimes called *special event tape*.

program monitor: The monitor on which the edits can be viewed as they are being made and on which the finished product appears.

publicity reel: The reel sent to television stations by publicists to introduce a person, an activity, or a product to the public. Celebrities who are changing their image will produce publicity reels, which will be disseminated as widely as possible.

rack focus: To manually change focus from a foreground object to a background object or vice versa.

rate card: A list of the services and equipment offered by a studio, postproduction house, or rental facility and the applicable rates charged.

rate of transition: The duration of a wipe, fade, or other transition—for example, a one-second dissolve.

raw stock: New tape that has never had an electronic signal recorded on it.

record machine: The three-quarter-inch, VHS, or 8-millimeter machine that records your material. The editing recorder is sometimes referred to as the *record machine* or as *record*.

red button: The tiny plastic button in all three-quarter-inch tapes that allows recording to occur. Remove this button to protect your material from being accidentally recorded over.

reel: A synonym for *demo tape*.

reel-to-reel: A system of tape transport used before cassette tapes became the standard. One-inch tape is still a reel-to-reel system, similar to reel-to-reel audio recorders.

resolution: The number of scan lines which a video system is capable of actually utilizing. Three-quarter-inch tape has 330 lines of resolution; S-VHS and Hi8 have 400; regular VHS has 240 to 260; the NTSC television standard has 525 (although not all lines are used for video); PAL has 600; household television sets range from 300 to 450; broadcast cameras and monitors can have as many as 600 or 700 lines of resolution.

sample reel: A synonym for *demo tape*.

saturation: The amount of color present on the video signal. Saturation may be increased or diminished by adjusting the chroma level on the time base corrector or processing amplifier.

scan lines: The lines that are electronically scanned to make up a television picture.

scew: A control that adjusts tape tension.

scratch tape: A tape used to record bits and pieces of material, such as photos or music, for transfer to the edited master. It is recycled after use.

scrim: See *diffusion screen*.

SECAM (séquential couleur à mémoire): A television standard created and used in France, among other countries. It is based on a different concept from NTSC and PAL.

SEG: See *special effects generator*.

setup: See *black level*.

short-load tape: Tapes loaded into cassettes in increments of five minutes. A four-minute program would be run on a five-minute tape.

shotgun microphone: A microphone with a very narrow (unidirectional) pickup pattern, which excludes most sound occurring on either side. Some shotguns can pick up sounds from directly behind them, however. These mikes are used when the desired audio takes place at a distance from the camera.

show number: Each episode of a show is numbered, and this number helps the production office locate the desired show. The number is found on the front page of episodic scripts.

signal-to-noise ratio: In any video picture, particularly those shot in low light, a certain amount of snow can be detected. This is referred to as noise, and the higher the signal-to-noise ratio, the less noise will be visible and the better the picture will be. In audio, it is the difference between the actual sound desired and the noise level on the tape. The wider the ratio, the better.

slave: In duplication, the tapes and recorders that are locked up to the master playback machine. They are said to be slaved to it.

slipped frames: One or two stray fields or frames of video that appear when the edit point has slipped, usually due to too many previews using control track. See also *flash frame*.

SLP: Acronym for *super long play*. See *extended long play*.

SMPTE time code: A system of permanently numbering each frame on a tape so it is always retrievable. Accepted as standard time code by the Society of Motion Picture and Television Engineers (SMPTE).

snow field: A brief section of blank tape created by the erase heads during assemble editing. Video cannot be inserted over a snow field.

soft cut: An edit that is a brief dissolve with a three-frame overlap on either side of the edit point. All television shows were once edited using soft cuts.

source deck: The machine that plays the material that will be edited.

source machine: See *source deck*.

source tape: The tape that contains the material that will be edited.

SP: See *standard play*.

special effects generator (SEG): A device that can switch between numerous video sources, such as cameras, videotape recorders, character generators, and graphics computers. It can key, can often chroma key, and produces a number of effects used when making transitions from one video source to another. Commonly called a *switcher*.

split screen: A screen that is split into two or more separate video pictures. A screen divided into four is known as a *quad*. Quads of four-camera sitcoms are often run (created), so the director and editor can watch all four angles simultaneously when making edit decisions.

spun Dacron: A type of diffusion material that is flame resistant. It replaces spun (fiber) glass as a diffuser.

standard play (SP): VHS tapes recorded at this speed can be edited directly in any studio with VHS editing equipment. Standard play is the fastest speed and yields the best picture quality.

standard speed: See *standard play*.

status monitor: The monitor that provides all the editing information to the editor. Among other things, it shows the hours, minutes, seconds, and frames for all tapes currently in the source and record machines.

straight-across transfer: A direct copy of an entire program from source directly to record with no edits in between.

straight cuts: When each new scene is cut directly onto the previous scene without fades, dissolves, or other transitions.

studio copy: A copy of a show that is kept for in-house viewing and can be borrowed to show to an agent or to lift a scene for an actor's reel or talk show clip. Sometimes called the *director's copy*.

submaster: To avoid re-editing an entire demo tape, the first edited master is copied down a generation to a submaster, where new scenes will be added or old scenes rearranged. It is a quick and inexpensive way to update a master by combining tape from the original edited master and new material from a source tape to use as a new current master and from which new copies will be made.

super: See *superimpose*.

superimpose (super): To put a name or titles over a background or other video.

super long play: See *extended long play*.

super VHS (S-VHS): An improved format offering 400 lines of resolution and other improvements in picture quality, gained by keeping the chrominance signal and luminance information separated in a component, rather than composite, system of transferring video information. This can result in a broadcast-quality picture, as with Hi8.

S-VHS: See *super VHS*.

switcher: See *special effects generator*.

switch on the fly: In a multicamera shoot or an A/B roll system, the shots are called as the show progresses in real time.

sync generator: A device that times all machines in an edit bay so that they are synchronized with one another. The sync generator sends its pulses to all the other machines in the system.

sync line: A single horizontal disturbance in the television picture that can be seen slowly rolling down the screen from top to bottom and beginning again at the top. It is caused by incorrect sync or an interruption of sync somewhere in the system. Once recorded, this condition cannot be corrected. The problem will not be apparent during E to E and is visible only during playback. To solve the problem, trace the incorrectly connected cable or wrongly placed sync switch and re-record the material from the point where the problem begins.

sync roll: Two or more source tapes from a multicamera shoot are rolled simultaneously or in sync, so that switching the fly can be done between them. This is the best of both worlds because previously recorded tapes can be studied and time can be saved by calling the shots without continuous stopping and starting.

syndication: A show that has finished its network run can be sold into syndication for airing at a local station's discretion.

tapestock: Blank videotape.

tape transport: The guides, rollers, and grooves that carry a tape around the video head drum.

TBC: See *time base corrector.*

Technowipe: A special lint-free cleaning cloth available in electronics supply stores.

three-chip camera: A high-end broadcast camera that uses three chips, one for each of the three colors on which color television is based: red, green, and blue. This provides the best possible picture quality. See *charged coupled device.*

three-quarter-inch tape: The industrial and archival standard for videotape. It may be slowly replaced by S-VHS or Hi-8 formats.

three-tube camera: See *three-chip camera.* Same as three-chip camera, but uses pickup tubes instead of chips.

time base corrector (TBC): A device that samples the unstable signal produced by the videotape and the videotape recorder, corrects it, and puts out a stable signal, which can then be combined with other video sources. It is sometimes capable of adjusting the video output and producing special effects, such as freeze frames, mosaics, and poster patterns.

time code: A system of permanently numbering each frame of video to facilitate scene retrieval and frame-accurate editing. The most common types are SMTE and CMX.

tracking: An adjustment that maximizes the scanner's ability to read the source tape. It is set automatically during record.

transfer: An individual copy of a tape. Also the act of making a copy. See also *copy, dub,* and *dupe.*

transition: One of various ways of moving from one scene to the next. Cuts, dissolves, and wipes are all transitions.

two-inch tape: An early videotape format that has generally been replaced by one-inch tape.

U-matic: A name for the three-quarter-inch cassette format that is based on the way the tape is threaded around the heads of the videotape recorder.

up-cut: To choose an earlier in point for the next edit in the scene just recorded.

upstream and downstream: A means of describing video to simplify communication. Everything is positioned from upstream to downstream in a video system. The source machine is upstream of the time base corrector, which is upstream of the switcher, which is upstream of the recorder, which is upstream of the program monitor. Likewise, the beginning of a show is upstream, and the end credits are downstream.

vanity tape: A tape made for an individual or family that contains material of personal interest, rather than material assembled solely for professional purposes.

vectorscope: A piece of equipment used to align and monitor the color (chrominance) in a video signal.

VHS tape: Half-inch tape originally designed for consumer use. It has become the corporate and industrial standard for the communication of information.

video heads: Tiny copper conductors that read the signal contained on the videotape or erase the signal, depending on their function.

video-only insert: A scene inserted within a previously recorded program that has a synchronous (clean) in point and out point and leaves the previously recorded audio intact.

video resume: An on-camera or video summation of work experience created to obtain employment.

viewing time: The rental of a video playback machine by the hour in order to review and log tapes before en edit session.

volume unit (VU) meter: A device used for monitoring the amplitude (volume) of the audio being recorded. See also *audio gain*.

VU meter: See *volume unit meter*.

waveform monitor (WFM): An oscilloscope that measures various elements of the video signal, including luminance (brightness) and black level.

WFM: See *waveform monitor*.

whip edit: See *wrong field edit*.

white balance: An adjustment used to render the camera's color interpretation accurately. To read colors correctly, the camera must be adjusted so that it will read a white image (usually a card or translucent lens cap) as white in the light where the recording will take place. Once white is balanced, all other colors will record correctly. See also *color temperature*.

window dub: A copy of source material on which time code numbers have been superimposed in the picture for making edit decisions.

wipe: A pattern used as a transition to open, close, or move from one sequence to another—for example, the old-time circle wipe where one scene emerges as a pinpoint from the center of another and eventually fills the screen.

work tape: See *scratch tape*.

wrong field edit: A black curve that suddenly appears in the picture at the edit point. It is technically known as a *wrong field edit* because it occurs when the machine tries to match field 1 with field 1 or field 2 with field 2. Also known as a *flag edit* or a *whip edit*.

zero decibels: The point on a volume unit meter at which the needles should most often peak for optimum sound reproduction.